Geographies of Entrepreneurship

T0265069

This book addresses a gap in the present literature on the role that geography plays in the distribution of entrepreneurial activity. Emerging work on entrepreneurial ecosystems suggests it is important entrepreneurship studies move beyond the mere identification of factors that impact entrepreneurial activity to consider the unique geographic contexts in which entrepreneurs operate. These contexts include a variety of interactive elements including regional characteristics, institutions, actors, and connectors.

As such, this collection analyses entrepreneurial activity in regions around the globe. The contributions explore a series of diverse regions in terms of their geographic, historical, industrial, and institutional contexts. The book also explores a range of topics, such as patterns of regional/subnational variations in entrepreneurial activity, geographically mediated determinants of entrepreneurship, inter-temporal dynamics, evolution of regional systems of entrepreneurship, and the impact of entrepreneurship on regional development and regional entrepreneurship policy. This book enhances our policy and practical knowledge about the unique regional context in which entrepreneurs operate and demonstrates the important role that geography plays in the spatial distribution of entrepreneurial activity.

Elizabeth A. Mack is Assistant Professor in the Department of Geography at Michigan State University, USA.

Haifeng Qian is Assistant Professor in the School of Urban and Regional Planning at the University of Iowa, USA.

Routledge Studies in Human Geography

This series provides a forum for innovative, vibrant, and critical debate within Human Geography. Titles will reflect the wealth of research which is taking place in this diverse and ever-expanding field. Contributions will be drawn from the main sub-disciplines and from innovative areas of work which have no particular sub-disciplinary allegiances.

For a complete list of titles in this series, please visit www.routledge.com.

Geographies of Entrepreneurship

Edited by
Elizabeth A. Mack and
Haifeng Qian

Routledge
Taylor & Francis Group

LONDON AND NEW YORK

First published 2016
by Routledge
2 Park Square, Milton Park, Abingdon, Oxon OX14 4RN

and by Routledge
711 Third Avenue, New York, NY 10017

First issued in paperback 2018

Routledge is an imprint of the Taylor & Francis Group, an informa business

British Library Cataloguing in Publication Data
A catalogue record for this book is available from the British Library

Library of Congress Cataloguing in Publication Data
Names: Mack, Elizabeth A., editor.
Title: Geographies of entrepreneurship / edited by Elizabeth A. Mack and Haifeng Qian. Description: Abingdon, Oxon; New York, NY : Routledge, 2016. | Series: Routledge studies in human geography Identifiers: LCCN 2015040926 | ISBN 9781138921061 (hardback) | ISBN 9781315686653 (e-book)
Subjects: LCSH: Economic geography. | Entrepreneurship. | Human geography–Economic aspects. Classification: LCC HF1025. G384 2016 | DDC 338/.04–dc23LC record available at http://lccn.loc.gov/2015040926

ISBN 13: 978-1-138-54647-9 (pbk)
ISBN 13: 978-1-138-92106-1 (hbk)

Typeset in Times New Roman
by Out of House Publishing

Contents

Figures

Tables

Contributors

Martin Andersson is a professor at CIRCLE, Lund University and the Department of Industrial Economics, Blekinge Institute of Technology.

Sandra Bürcher is a PhD student in economic geography at the Institute of Geography of the University of Bern. Her thesis focuses on economic development of peripheral regions from an embeddedness perspective.

Roberta Capello is Professor of Regional Economics in the Department of Architecture, Built Environment and Construction Engineering (ABC), Politecnico di Milano.

Kevin Credit is a PhD student in geography at Michigan State. The overall goal of his research is to better understand how planning (broadly defined) and urban form influence entrepreneurship, innovation, and economic diversity. Kevin's recent and ongoing dissertation work looks at the impact of public transit on the spatial distribution of new businesses, as well as the role of transportation factors in entrepreneurial location decisions.

Michael Fritsch is Full Professor of Economics and the Chair of Business Dynamics, Innovation, and Economic Change at the Friedrich-Schiller-University Jena, Germany. He is also a research professor at the German Institute of Economic Research (DIW Berlin) and at the Halle Institute for Economic Reseach (IWH). He is an Associate Editor of the academic journals *Regional Studies* and *Small Business Economics*. The main fields of his recent research are entrepreneurship and innovation.

Qi Guo is a PhD student at the College of Urban and Environmental Sciences, Peking University.

Antoine Habersetzer is a PhD student in economic geography at the University of Bern. In his dissertation, he investigates spinoff dynamics and entrepreneurial heritage of manufacturing firms in Swiss peripheral regions.

Canfei He is Professor of Economic Gepgraphy at the College of Urban and Environmental Sciences, Peking University.

Jun Koo is a professor in the Department of Public Administration at Korea University. He received his PhD in city and regional planning from the University of North Carolina at Chapel Hill. He has diverse research interests, including regional development, innovation, entrepreneurship, and happiness. His research has appeared in many international journals and books in regional science, urban planning, business, and public administration.

Sierdjan Koster is an assistant professor in the economic geography department of the Faculty of Spatial Sciences, part of the University of Groningen in the Netherlands. In his research, he studies the relationship between the labor market careers of entrepreneurs, the characteristics of the firms they start and the performance and wider regional economic impact of such firms. Currently, he also is an editor for *REGION*, the online and open access journal of the European Regional Science Association.

Niclas Lavesson is a PhD student in economic geography at CIRCLE and Department of Human Geography, Lund University.

Camilla Lenzi is Associate Professor of Applied Economics in the Department of Architecture, Built Environment and Construction Engineering (ABC), Politecnico di Milano.

Elizabeth A. Mack is Assistant Professor in the Department of Geography at Michigan State University where she teaches courses in economic geography. She is also a research affiliate of the GeoDa Center for Geospatial Analysis and Computation. Recently, Dr. Mack's work is focused on understanding entrepreneurial ecosystems and the enabling role of broadband Internet connections in the new venture creation process.

Heike Mayer is Professor of Economic Geography in the Institute of Geography and co-director of the Center for Regional Economic Development at the University of Bern. Her primary area of research is in local and regional economic development with a particular focus on entrepreneurship and innovation dynamics.

Haifeng Qian is Assistant Professor in the School of Urban & Regional Planning, University of Iowa. His research interests include regional economic development, entrepreneurship and innovation, and policy analysis. He is an associate editor of *Economic Development Quarterly*.

Ben Spigel is a lecturer and Chancellor's Fellow in the Entrepreneurship and Innovation Group at the University of Edinburgh Business School.

Michael Wyrwich is Research Associate at the Chair of Business Dynamics, Innovation, and Economic Change at the Friedrich-Schiller-University Jena, Germany. His main research interests are focused institutional and cultural aspects of regional entrepreneurship as well as persistence in eco-nomic development.

Songhee Yoo is a doctoral candidate in the Department of Public Administration at Korea University. She has keen interest in regional development, regulation, and public finance.

Shengjun Zhu is a lecturer in the Department of Geography, College of Science, Swansea University.

Preface

This book is a product of sessions related to research in the geography of entrepreneurship that were initiated at the Association of American Geographers (AAG) meetings in 2012 and that continue to this day. The goal of these sessions was to bring to together researchers who were doing work that addressed the geographical context in which entrepreneurship takes place, as well as regional specific factors that explain the uneven distribution of entrepreneurial activity over space between and within regions.

List of Past AAG Sessions:

2012: Entrepreneurship, Innovation, and Regional Prosperity: Creating Competitive Economies (AAG New York, NY)

2013: Geographies of Entrepreneurship in Developed and Developing Economies (AAG Los Angeles, CA)

2014: Geographies of Entrepreneurship (AAG Tampa, FL)

2015: Geographic Perspectives on Entrepreneurship: Factors, Processes, and Disparities (AAG Chicago, IL)

Another goal of the AAG sessions listed above was to encourage geographically oriented research in entrepreneurship, which to date, has widespread involvement from numerous disciplines including: management, international business, economics, sociology, and planning. To a large extent, geographers have not engaged in this field of research as readily as have these other disciplines. Edward Malecki of the Ohio State University is an exception to this trend and has been one of the few geographers engaged in entrepreneurship for some time. This is unfortunate because the spatial perspective and analytical toolkit of the geographer have the potential to shed light on what we know to be an embedded local event.

As the world flattens and innovation transpires at an increasingly rapid pace, the value of entrepreneurs and small business is bound to grow. In this environment, the nimble innovativeness of entrepreneurs will help their businesses, and the economies of the regions in which they operate, adapt to the lockstep forces of globalization and technological change.

This book should appeal to a wide range of scholars, practitioners, and policymakers who are passionate about entrepreneurship. We hope you enjoy reading this book and perhaps even decide to join us and the contributors to this book in our journey to unravel the spatial dimensions of entrepreneurial activity.

Elizabeth A. Mack and Haifeng Qian
September 2015

Acknowledgements

Many colleagues, scholars, policymakers, and students have served as sources of inspiration that precipitated the compilation of this edited volume. Our thanks to these folks and entrepreneurship scholars all over the world. While we cannot hope to acknowledge each of these folks individually, we would like to express our thanks to the people and organizations that were instrumental in the production of this book. Thank you to the Association of American Geographers (AAG) for supporting these sessions at the annual conference and to all the people that helped organize sessions at this conference each year: Kevin Stolarick in 2012, Ben Spigel in 2013, and Yasuyuki Motoyama in 2014 and 2015. Thank you to Yasuyuki Motoyama and the Ewing Marion Kauffman Foundation for Entrepreneurship for their participation and support of the AAG sessions every year. Thank you to AAG session participants past, present, and future for sharing your ideas and your commitment to unraveling the factors and processes that drive entrepreneurship and their variation over space and through time. A final thanks goes out to entrepreneurs around the globe. You are a critical component to the vitality of regional economies, and your innovative spirit and determination have transformed the development trajectory of many places in a positive manner.

1 The geography of entrepreneurship

Elizabeth A. Mack

Introduction

Entrepreneurial activity, like many phenomena, is unevenly distributed across geographic space. The entrepreneurial landscape is dotted with well-known hubs of activity such as Silicon Valley and Route 128 in Boston (Saxenian, 1994), as well as the Third Italy and the Sakaki district in Japan (Malecki, 1993). This same landscape is also made up of many lagging regions in deindustrialized and remote locales that are on the periphery of the global entrepreneurship community. It is these two extremes, and the increasingly important role of entrepreneurs and small businesses in a dynamic, informational economy, that drives research attempts to explain why some places are more entrepreneurial than others. This drive stems from the many benefits entrepreneurs bestow on their communities, including the creation of jobs (Kirchhoff and Phillips, 1988; Von Bargen et al., 2003; Baptista et al., 2008) and the diversification of the industrial base (Boschma and Frenken, 2006; Andersson and Koster, 2011). Unpacking the entrepreneurial landscape is a rather complex task however, because of the embeddedness of entrepreneurs in their local environments (Jack and Anderson, 2002; Welter and Smallbone, 2011). This embedded nature means that the individual characteristics of entrepreneurs, as well as their recognition and pursuit of opportunities, are moderated by the unique legal, institutional and social characteristics of regions (Audretsch, 2007). This blend of regional characteristics comprises the milieu of entrepreneurial communities around the world.

Regional determinants of entrepreneurial activity

A lot of progress has been made in understanding the factors that explain the uneven distribution of entrepreneurial activity across space. Work on the demography of the regional population highlights the role that gender (Langowitz and Minnitti, 2007), age (Curran and Blackburn, 2001), race (Butler and Herring, 1991; Butler, 2012) and ethnicity (Wilson and Martin, 1982) play in the entrepreneurial propensity of the local population. These characteristics, as well as the educational attainment of the local labor force, are important to consider because of the propensity for entrepreneurs to start businesses near home (Figueiredo et al., 2002; Stam, 2007), and for businesses to have higher survival rates when started near entrepreneurs' homes (Dahl and Sorenson, 2012). Thus, research suggests that locales with the right demographic mix of highly educated people with the proper industry experience (Shane and Stuart, 2002) are likely to be more entrepreneurial than others.

Aside from population characteristics, the mix of existing businesses within regional environments has an impact on local entrepreneurial activity. Firm size (Armington and Acs, 2002), as well as the level of regional specialization (Acs and Armington, 2004; Campi et al., 2004) are noted to have an impact on entrepreneurial activity. Studies have found for example that industrial diversity is more conducive to entrepreneurial activity than is regional specialization (Glaeser et al., 1992; Feldman and Audretsch, 1999; Acs and Armington, 2004). This argument is rooted in Jane Jacobs' externalities (Jacobs, 1969), which facilitate the exchange of complementary knowledge across a diverse array of economic actors, which leads to the production of new knowledge (Audretsch and Thurik, 2004).

In addition to the existing mix of people and businesses, studies have found that the quality of networking activity in regional environments is an important component of vibrant entrepreneurial milieu. This is perhaps one of the more important cultural factors that distinguishes Silicon Valley from other places (Saxenian, 1994). In their contribution to this book in Chapter 4, Michael Fritsch and Michael Wyrwich note that regional cultures are slow to change, which produces path dependency in the trajectory of entrepreneurial activity over time. In this respect, there are cumulative causation elements to entrepreneurial activity (Holcombe, 2007); more historical activity breeds more activity in the future.

Entrepreneurial ecosystems

Concerns about the entrepreneurial inertia of places, combined with the recognition that entrepreneurship is a highly contextual local event (Feldman, 2001), have prompted researchers to delve more deeply into the regional profiles of place characteristics from a systems perspective. This systems perspective uses the term "entrepreneurial ecosystem" to group and organize the set of interacting components within regions that foster entrepreneurial activity (Neck et al., 2004). Examples of successful ecosystems around the world include the Copenhagen pharmaceuticals cluster (Mason and Brown, 2014), and Oxford, England (Lawton Smith et al., 2008; Mason and Brown, 2014). Research on entrepreneurial ecosystems take perhaps a more comprehensive approach to documenting the unique regional contexts in which entrepreneurs operate and work, and this newer area of entrepreneurship research is growing rapidly. To date, the bulk of work has been dedicated to documenting the components of ecosystems (Bahrami and Evans, 1995;

Cohen, 2006; Isenberg, 2010, 2011). Some work has begun to expand on this documentation approach to consider evolutionary perspectives on ecosystem dynamics (Mack and Mayer, 2015), but much more comparative work on the entrepreneurial ecosystem is needed. This research need and future directions for entrepreneurship research will be discussed in the closing chapter of this book.

Spatial approaches to entrepreneurship?

Despite these advances in entrepreneurship, much more work is required to unpack the spatialities of the actors, factors, and processes that foster entrepreneurship. While case study work (Saxenian, 1994; Feld, 2012; Spigel, 2015) has provided valuable research about particular places, more comparative work is needed, as is research that evaluates the variability in ecosystem components over space and time. These types of research areas require the tools and techniques of the geographer to unravel the role that space and place play in the distribution of entrepreneurs, and the resources they require to start successful enterprises.

This is not to say that the role of space has been completely ignored in prior research efforts. For some time now, scholars have recognized the spatial dimension to entrepreneurial activity (Acs and Armington, 2006; Renski, 2008; Glaeser and Kerr, 2009; Plummer and Pe'er, 2010). Ten years ago, for example, Andersson (2005) outlined how a spatial perspective can help us understand the role of entrepreneurship in economic development by providing greater resolution on the distribution of profit-making opportunities, knowledge transmission, and variations in culture and institutions. Although research in entrepreneurship is expanding rapidly, a comparatively small percentage of this research considers *explicitly* the spatial dimensions of entrepreneurial activity. Instead, the spatial dimension of the factors and processes that are responsible for the uneven distribution of entrepreneurial activity are often subsumed in theoretical and quantitative studies. Plummer and Pe'er (2010) argue that the voluminous research on geography and entrepreneurship falls primarily into the category of spatial economics, which take an ironically aspatial approach to research.

Thus, although there appears to be a wide body of literature on the "geography of entrepreneurship," only a very small portion of this literature examines geography as a fundamental factor in the distribution of the people and factors that promote entrepreneurship. This is an unfortunate oversight if we are to understand why some places are more entrepreneurial than others. This is not to say that spatial work is completely absent from entrepreneurship research. In fact, work by Audretsch and Fritsch (1994), Acs and Armington (2006), Glaeser and Kerr (2009) and Renski (2008) provides a more geographical perspective on entrepreneurship using the tools of the geographer (i.e. maps). As with research on the entrepreneurial ecosystem, however, much more work remains to unravel the geographic locations within countries, regions, and cities that have an overrepresentation of entrepreneurial activity. Increasingly, the region as the nexus of analysis is becoming even more valuable as a unit of analysis (Scott, 1998). Therefore, aggregate analyses of entrepreneurship that ignore regional context hold less value than they did previously; particularly if we are trying to understand the best manner to foster entrepreneurship in vastly different locales.

Content of book

Given this challenge, the present edited volume is an effort to initiate an in-depth conversation about the role of geography in entrepreneurship. Contributors for this book were asked to take a geographical perspective on entrepreneurship to consider unique trends and issues that impact the incidence of entrepreneurial activity in countries around the globe. The perspectives provided in this edited volume are geographic in two ways: (1) they are from select countries around the globe; and (2) they consider a range of issues that impact the spatial distribution of entrepreneurial activity within a particular country. Contributions for this book were solicited from researchers around the globe to gain a diversity of perspective from a range of regions and countries. The opening chapter of the book tackles the issue of innovation in European regions. The middle six chapters of the book cover entrepreneurial issues in six different countries: Germany, Sweden, China, South Korea, the United States and Canada. The last chapter of the book discusses entrepreneurship in peripheral regions, which is important to consider given the emphasis of research on entrepreneurship in urban locales.

Chapter 2, by Roberta Capello and Camilla Lenzi, revisits the ideas of Schumpeter (1934), who first suggested the linkage between innovation and entrepreneurial opportunity. In particular, the chapter provides greater resolution on the innovation–entrepreneurship nexus by analyzing the spatial variation in innovation across European regions or "regional patterns of innovation." The chapter classifies regions based on three archetypes of innovation: a science-based pattern, a creative application pattern and an imitative pattern. These three archetypes may be expanded upon to highlight five types of European regions based on three key attributes:

1   EU share of all patents;
2   share of all EU firms that introduce a product and/or process innovation; and
3   the share of firms that introduce marketing and/or organizational innovations.

The five types of regions identified are:

1   European science-based areas;
2   applied science areas;

3   smart technology areas;
4   smart and creative diversification areas; and
5   imitative innovation areas.

Entrepreneurial activity is then examined in these types of regions based on an index of regional entrepreneurship called REDI (Regional Entrepreneurship and Development Index). Finally, the coincidence between entrepreneurship, as measured by REDI, and innovation is analyzed. This analysis reveals the link between potential conditions for entrepreneurship and the innovation capacity of regions; the richer the potential conditions for entrepreneurship, the more advanced is the type of innovation model within the region. These results have important implications for regional innovation policy, particularly those founded on smart specialization. They suggest that policy interventions used to strengthen and diversify regional assets and capacities must also consider the innovation environment in which entrepreneurs operate, and the varied strength of the dimensions (opportunity perception, risk orientation and strategic vision) to entrepreneurial activity in each region.

Chapter 3 by Ben Spigel provides a more fine-grained perspective on entrepreneurial networks outside of Silicon Valley and examines the networking practices of Canadian entrepreneurs in Waterloo, Ontario and Calgary, Alberta through the lens of a Bourdieuian framework. The goal of the analysis is to highlight the cultural embeddedness of these networking practices given the spatio-temporal variation in these interactions. A Bourdieuian framework is useful for understanding the institutional context of networking since this is the product of the interaction between entrepreneurs' daily routines, the rules, norms and hierarchies of a place, and the interpretation of this institutional context by each entrepreneur. Technology entrepreneurs are the focus of the piece given their balancing act between two fields, a technology entrepreneurship field (FTE) and a local field. The chapter argues that what people perceive as regional entrepreneurial culture is the outcome of a series of choices made by individuals about the best practices for doing business. The results of this comparative analysis highlight different networking practices for entrepreneurs in Waterloo and Calgary. In Waterloo, entrepreneurs are more likely to network with one another to enhance their business skills and build social capital in the entrepreneurship community. In Calgary, however, entrepreneurs concentrate their networking efforts on potential clients, and spend little time networking with other entrepreneurs given the for-profit orientation of the local culture. These results highlight that entrepreneurship research should focus on networking practices rather than the description of network characteristics such as size, density, and connectedness. From a practical perspective, efforts to foster a "regional culture" of entrepreneurship must have a clear understanding of the fields that entrepreneurs navigate, and then work within these fields to help entrepreneurs grow their businesses. Imposing a formula for building regional culture that is transplanted from other locales should be avoided.

Chapter 4, by Michael Fritsch and Michael Wyrwich, attempts to shed light on the reasons behind the persistent nature of entrepreneurial activity over time (Fritsch and Mueller, 2007; Andersson and Koster, 2011). It examines the impact of different types of self-employment on the persistence of entrepreneurship for regions in West Germany. Although the authors suspect that persistence is grounded in the immobility of regional factors that foster entrepreneurship, the chapter argues that the sources of persistence are actually unclear to this date. Using information about different types of self-employment including: non-agricultural private sector employment, self-employment in knowledge-intensive industries, female and male self-employment, and historical information about homeworkers; econometric models are estimated to test whether these specific types of self-employment positively impact entrepreneurial persistence over time. Entrepreneurship in the chapter is defined as the average start-up rate between 1976 and 2010. Results of the analysis highlight that in the West German case, the type of self-employment considered has a very different relationship with the persistence of entrepreneurship over time. Specifically, types of self-employment that proxy for non-necessity types of entrepreneurial activity (self-employment in knowledge-intensive industries as well as male self-employment in non-agricultural industries) have the greatest impact on persistence over time. Although it is likely that the persistence of entrepreneurship will vary according to regional context, analytical results suggest important lessons for creating regional policies to foster entrepreneurship. First, fostering new venture creation in sectors with the most impact on activity is likely to be more efficient than that concentrating very broadly on any type of new venture. Second, policies designed to foster new venture creation should be designed with a long-term orientation in mind.

Chapter 5, by Jun Koo and Songhee Yoo, provides an in-depth look at the geographic distribution of high-technology ventures and necessity-based entrepreneurship (traditional small business) in South Korea. It also looks at central and local policies for fostering entrepreneurial activity in each of these types of new business. Results of the analysis highlight that technology businesses are highly concentrated in the capital region of the country. Despite the perception that entrepreneurship in South Korea is high-technology oriented, this is not the case: high-technology ventures account for just 2 percent of all entrepreneurial activities. This perception is likely due to broad-based definitions of technology-oriented firms by the Korean government, which include more traditional manufacturing sectors such as paper products. In fact, the study highlights that global players are the real source of technology innovation in the country; smaller, locally based companies are not the main source of technology innovation. Given these results, the authors suggest there is a policy mismatch in the country because of the emphasis of central and local government policies on technology ventures. This is not only a concern given the large number of non-technology or traditional enterprises that make up all new ventures, but the low survival rates of traditional new ventures. Anecdotal evidence from interviews with Korean entrepreneurs suggests some revisions to address this policy mismatch. For example, changes are necessary to reduce the amount of time required for high-tech ventures to receive seed funding, and to reduce the excessive paperwork required of entrepreneurs in South Korea. Traditional small business owners also highlighted a lack of knowledge about support programs and opportunities available to them.

Chapter 6, by Canfei He, Qi Guo and Shengjun Zhu, examines the institutional context in which entrepreneurial activity is cultivated in China. Over time, this context has evolved rapidly since the Communist Party took over power in 1949. After this time, policies designed to reform and open the economy have submitted entrepreneurs to the forces of marketization, globalization and decentralization. The authors define marketization as the change from a collective-state owned system to one where private ownership is more prevalent. Decentralization refers to the move from a centrally planned orientation to one where regional and local governments are more involved in politics and economic development efforts. The force of globalization refers to the opening up of the economy to trade and foreign direct investments (FDI). This chapter uses data from the Annual Survey of Industrial Firms (ASIF) to examine spatio-temporal patterns in the levels of Chinese entrepreneurship between 1998 and 2008, and analyze how the three forces of marketization, globalization and decentralization have impacted entrepreneurship. In this ten-year period, the analysis highlights that entrepreneurship is an increasingly important component of the Chinese economy because privately owned start-ups make up an increasingly large portion of entrepreneurial activity. The size of start-ups is also shrinking, and these start-ups appear to have more of a domestic rather than an export-orientation. There are distinct spatial variations in entrepreneurship that exhibit a declining gradient that is highest in the eastern portion of the country and lowest in the west. Variations in activity between provinces have also declined over time, but increased within provinces. Analytical results highlight that locales in the country with a global, market orientation are more conducive to entrepreneurial activity. Cities with protection policies are less conducive to entrepreneurial activity compared to those that offer subsidies.

Chapter 7, by Elizabeth A. Mack and Kevin Credit, explores spatial, temporal and industrial variations of new business activity for ten metropolitan areas in the United States between 1989 and 2010. This chapter is an effort to provide more spatial resolution on entrepreneurial activity within metropolitan areas given the emphasis of prior work on aggregate levels of analysis such as labor market areas (LMAs) (Acs et al., 2007). Given the role of entrepreneurs as agents of evolutionary economic change (Stam and Lambooy, 2012) it is important to understand where new firm activity is located within urban environments to better inform urban revitalization efforts, as well as initiatives to diversify the regional industrial base. Using a point-level dataset on new firm activity, the chapter finds distinctive patterns in new establishment starts. New firm activity has increased steadily over this 21-year period and is also strongly associated with national business cycles. From a spatial perspective, establishment starts have intensified in urban centers and also followed the suburban and exurban movement of people from central city locations. New business starts in this 21-year period also have a distinctive retail and services industrial orientation. An interesting aspect of the services starts is their concentration in packaging and labeling services as well as trade show and convention organization. This industrial orientation could be a result of advances in Internet-related information and communications technologies (ICTs) that enable online shopping, which then spurs demand for packaging and shipping services. It also likely reflects the increased incidence of temporary clustering opportunities at trade fairs and conventions (Maskell et al., 2006). In this sense, the intra-metropolitan location of new business trends mirrors the movement patterns of people within cities as well as the post-industrial orientation of the global economy.

Chapter 8, by Martin Andersson, Sierdjan Koster and Niclas Lavesson, explores differences in start-up characteristics across the urban hierarchy in Sweden. This exploratory study is designed to fill the gap in prior work which emphasizes the frequency rather than the underlying characteristics of start-ups. Work that fills this gap is important because it will shed light on varying regional impacts of start-ups on job creation, which is likely related to the type rather than the number of start-ups. The characteristics of start-ups examined in this chapter include the labor market background of the entrepreneur, as well as the size, education level of firm employees and industry membership of the firm. Four types of start-ups are analyzed to characterize differences in the quality of start-ups: the self-employed, spinoffs, new firms created by entrepreneurs that were unemployed previously, and "other" firms. An urban hierarchy is constructed for Sweden using municipal level data. The Stockholm, Göteborg and Malmö regions sit at the top of the urban–rural Swedish hierarchy. Conversely, the countryside contains the most municipalities (164) but accounts for a fairly low percentage of both population and employment (30 percent). Results of the analysis highlight that although self-employment comprises the majority of start-ups, there is little difference in the distribution of the four types of start-ups across the urban–rural hierarchy. Start-ups in central locales are larger than those in rural locales. Central locations are also more likely to have a greater amount of knowledge-intensive business services. Less knowledge-intensive and high-tech manufacturing start-ups exhibit a somewhat flatter distribution across this hierarchy with no real differences between urban and rural areas. Start-ups in urban areas have a more highly educated workforce than do rural start-ups. Combined, these results suggest spatial sorting of highly educated entrepreneurs and workers to urban locales. In this regard, more work is need to understand spatial variations in the characteristics of start-ups. This includes investigations of this relationship in other countries and economic contexts.

Chapter 9, by Sandra Bürcher, Antoine Habersetzer and Heike Mayer, discusses entrepreneurship in peripheral regions. This is an important chapter to include because not all entrepreneurial activity thrives in dense urban environments. In fact, in some urban environments the negative externalities associated with urbanity (crime, pollution and congestion) exceed the marginal benefits of central, urban locations (Stam and Lambooy, 2012). The chapter begins by emphasizing that peripheral regions are not the same, and may be considered peripheral in a variety of contexts. These diverse contexts include geographic distance from core regions, as well as organizational distance or dissimilarity. From a regional innovation systems (RIS) perspective, regions may be peripheral due to poor absorptive capacity, the absence of agglomeration economies, a lack of support organizations, and/or lower degrees of connectedness with other places. The authors argue that the heterogeneous nature of peripheral places, which actually comprises a spectrum of peripherality, merits a more nuanced perspective on entrepreneurship in peripheral places. To provide more insights into these locales, this chapter offers a relational perspective of peripheral regions constructed from the entrepreneurial heritage and embeddedness approaches. The discussion of entrepreneurship in peripheral places from these perspectives highlights both potential advantages and disadvantages of these locales, which culminates with a typology of peripheral regions. This typology highlights the need for detailed documentation of the problems and prospects of peripheral regions in order to construct region-specific solutions to unfavorable entrepreneurial milieux.

Combined, these chapters highlight the value of a geographic perspective in addressing key themes in entrepreneurship research. These themes include the distribution of innovative activity over space in Chapter 2. The geographic and cultural context of networking activity in locales outside of Silicon Valley in Chapter 3 and spatial inertia or entrepreneurial persistence in Chapter 4. The remaining five chapters analyze variations in the spatial distribution of new ventures over space in different geographic, institutional, industrial and temporal contexts. These themes are categorized into two overarching themes in the concluding chapter: the role of entrepreneurship in regional development, and geographical factors impacting regional variations in new venture activity. From these two broad themes, three lines of future research that are discussed at greater length in the conclusion: the geography of knowledge spillovers, entrepreneurial ecosystems and evolutionary economic geographic perspectives on entrepreneurship. Although this book is not able to address all of the ways in which an *explicitly* geographic perspective adds value to many cross-cutting themes in entrepreneurship, we hope that *Geographies of Entrepreneurship* encourages a broad range of spatial inquiry into a number of research themes in entrepreneurship. This geographic approach is absolutely vital to advancing our understanding of the evolving position of a diverse range of entrepreneurial spaces around the world.

References

Acs, Z. and Armington, C. (2004). Employment growth and entrepreneurial activity in cities. *Regional Studies*, 38(8), 911–927.

Acs, Z. J. and Armington, C. (2006). *Entrepreneurship, geography, and American economic growth*. Cambridge: Cambridge University Press.

Acs, Z. J., Armington, C. and Zhang, T. (2007). The determinants of new-firm survival across regional economies: The role of human capital stock and knowledge spillover. *Papers in Regional Science*, 86(3), 367–391.

Andersson, E.M. (2005). The spatial nature of entrepreneurship. *The Quarterly Journal of Austrian Economics*, 8(2), 21–34.

Andersson, M. & Koster, S. (2011). Sources of persistence in regional start-up rates: Evidence from Sweden. *Journal of Economic Geography*, 11(1), 179–201.

Armington, C. and Acs, Z. J. (2002). The determinants of regional variation in new firm formation. *Regional Studies*, 36(1), 33–45.

Audretsch, D. B. (2007). Entrepreneurship capital and economic growth. *Oxford Review of Economic Policy*, 23(1), 63–78.

Audretsch, D. B. and Fritsch, M. (1994). The geography of firm births in Germany. *Regional Studies*, 28(4), 359–365.

Audretsch, D. B. and Thurik, A. R. (2004). *A model of the entrepreneurial economy* (No. 1204). Papers on entrepreneurship, growth and public policy.

Bahrami, H. and Evans, S. (1995). Flexible re-cycling and high-technology entrepreneurship. *California Management Review*, 37(3), 62–89.

Baptista, R., Escária, V. and Madruga, P. (2008). Entrepreneurship, regional development and job creation: The case of Portugal. *Small Business Economics*, 30(1), 49–58.

Boschma, R. A. and Frenken, K. (2006). Why is economic geography not an evolutionary science? Towards an evolutionary economic geography. *Journal of Economic Geography*, 6(3), 273–302.

Butler, J. S. (2012). *Entrepreneurship and self-help among black Americans: A reconsideration of race and economics*. Albany: SUNY Press.

Butler, J. S. and Herring, C. (1991). Ethnicity and entrepreneurship in America: Toward an explanation of racial and ethnic group variations in self-employment. *Sociological Perspectives*, 34(1), 79–94.

Campi, M. T. C., Blasco, A. S. and Marsal, E. V. (2004). The location of new firms and the life cycle of industries. *Small Business Economics*, 22(3–4), 265–281.

Cohen, B. (2006). Sustainable valley entrepreneurial ecosystems. *Business Strategy and the Environment*, 15, 1–14.

Curran, J. and Blackburn, R. A. (2001). Older people and the enterprise society: Age and self-employment propensities. *Work, Employment & Society*, 15(4), 889–902.

Dahl, M. S. and Sorenson, O. (2012). Home sweet home: Entreprenuers' location choices and the performance of their ventures. *Management Science*, 58(6), 1059–1071.

Feld, B. (2012). *Startup communities: Building an entrepreneurial ecosystem in your city*. Hoboken: John Wiley & Sons.

Feldman, M. P. (2001). The entrepreneurial event revisited: Firm formation in a regional context. *Industrial and Corporate Change*, 10(4), 861–891.

Feldman, M. P. and Audretsch, D. B. (1999). Innovation in cities: Science-based diversity, specialization and localized competition. *European Economic Review*, 43(2), 409–429.

Figueiredo, O., Guimaraes, P. and Woodward, D. (2002). Home-field advantage: Location decisions of Portuguese entrepreneurs. *Journal of Urban Economics*, 52, 341–361.

Fritsch, M. and Mueller, P. (2007). The persistence of regional new business formation activity over time: Assessing the potential of policy promotion programs. *Journal of Evolutionary Economics*, 17(3), 299–315.

Glaeser, E. L. and Kerr, W. R. (2009). Local industrial conditions and entrepreneurship: how much of the spatial distribution can we explain? *Journal of Economics & Management Strategy*, 18(3), 623–663.

Glaeser, E., Kallal, H., Sheinkman, J. and Schleifer, A. (1992). Growth in cities. *Journal of Political Economy*, 100, 1126–1152.

Holcombe, R. G. (2007). Entrepreneurship and economic growth. In: Powell, B. (ed.), *Making poor nations rich: Entrepreneurship and the process of economic development*. Stanford: Stanford University Press.

Isenberg, D. J. (2010). How to start an entrepreneurial revolution. *Harvard Business Review*, 88(6), 40–50.

Isenberg, D. (2011). *The entrepreneurship ecosystem strategy as a new paradigm for economic policy: Principles for cultivating entrepreneurship*. Dublin: Institute of International European Affairs.

Jack, S. L. and Anderson, A. R. (2002). The effects of embeddedness on the entrepreneurial process. *Journal of Business Venturing*, 17(5), 467–487.

Jacobs, J. (1969). *The economy of cities*. New York: Random House.

Kirchhoff, B. A. and Phillips, B. D. (1988). The effect of firm formation and growth on job creation in the United States. *Journal of Business Venturing*, 3(4), 261–272.

Langowitz, N. and Minniti, M. (2007). The entrepreneurial propensity of women. *Entrepreneurship Theory and Practice*, 31(3), 341–364.

Lawton-Smith, H. L., Romeo, S. and Bagchi-Sen, S. (2008). Oxfordshire biomedical university spin-offs: An evolving system. *Cambridge Journal of Regions, Economy and Society*, 1(2), 303–319.

Mack, E. A. and Mayer, H. (2015). The evolutionary dynamics of entrepreneurial ecosystems. *Urban Studies*. DOI: 10.1177/0042098015586547.

Malecki, E. J. (1993). Entrepreneurship in regional and local development. *International Regional Science Review*, 16(1–2), 119–153.

Maskell, P., Bathelt, H. and Malmberg, A. (2006). Building global knowledge pipelines: The role of temporary clusters. *European Planning Studies*, 14(8), 997–1013.

Mason, C. and Brown, R. (2014). Entrepreneurial ecosystems and growth oriented entrepreneurship. Retrieved September 10, 2014, from www.oecd.org/cfe/leed/Entrepreneurial-ecosystems.pdf.

Neck, H. M., Meyer, G. D., Cohen, B. and Corbett, A. C. (2004). An entrepreneurial system view of new venture creation. *Journal of Small Business Management*, 42(2), 190–208.

Plummer, L. A. and Pe'er, A. (2010). The geography of entrepreneurship. In: Acs, Z. J. and Audretsch, D. B. (eds.), *Handbook of entrepreneurship research* (pp. 519–556). New York: Springer.

Renski, H. (2008). New firm entry, survival, and growth in the United States: A comparison of urban, suburban, and rural areas. *Journal of the American Planning Association*, 75(1), 60–77.

Saxenian, A. (1994). *Regional advantage: Culture and competition in Silicon Valley and Route 128*. Cambridge, MA: Harvard University.

Schumpeter, J. A. (1934). *The theory of economic development: An inquiry into profits, capital, credit, interest, and the business cycle*. Cambridge, MA: Transaction Publishers.

Scott, A. J. (1998). *Regions and the world economy*. Oxford: Oxford University Press.

Shane, S. and Stuart, T. (2002). Organizational endowments and the performance of university start-ups. *Management Science*, 48(1), 154–170.

Spigel, B. (2015). The relational organization of entrepreneurial ecosystems. *Entrepreneurship Theory and Practice*, DOI: 10.1111/etap.12167.

Stam, E. (2007). Why butterflies don't leave: Locational behavior of entrepreneurial firms. *Economic Geography*, 83(1), 27–50.

Stam, E. and Lambooy, J. (2012). Entrepreneurship, knowledge, space, and place: Evolutionary economic geography meets Austrian economics. In: Andersson, D. E. (ed.), *The spatial market process* (pp. 81–103). Bingley: Emerald Group Publishing.

Von Bargen, P., Freedman, D. and Pages, E. R. (2003). The rise of the entrepreneurial society. *Economic Development Quarterly*, 17(4), 315–324.

Welter, F. and Smallbone, D. (2011). Institutional perspectives on entrepreneurial behavior in challenging environments. *Journal of Small Business Management*, 49(1), 107–125.

Wilson, K. L. and Martin, W. A. (1982). Ethnic enclaves: A comparison of the Cuban and Black economies in Miami. *American Journal of Sociology*, 88(1), 135–160.

2 The geography of the innovation–entrepreneurship nexus in Europe

Roberta Capello and Camilla Lenzi

Introduction

Following the seminal works by Schumpeter (1934) and Kirzner (1973), and more generally of the Austrian school, entrepreneurship has always been considered a constituent part of any innovative process and, by extension, a crucial determinant of economic performance.

The abundance of studies that have recently flourished, however, has not yet translated into a consistent body of results; still, empirical evidence is mixed and a clear and undisputable confirmation of the impact of entrepreneurship on regional performance is still missing. Rather, the empirical evidence increasingly questions the existence of a direct and automatic link between new firms creation and economic performance; this link, instead, can vary in significance, intensity, sign and time to take place according to both industry-specific conditions and the regional environment (Fritsch and Storey, 2014).

In this chapter, we propose that a possible lens to read this growing but often incomparable, conflicting and, sometimes, 'stand-alone' pieces of evidence is to take a step back. Rather than looking directly at the relationship between entrepreneurship and regional performance, there is still need to better understand the interplay between innovation and entrepreneurship in space, so to underline in which situations entrepreneurship is actually accompanied with innovation and can therefore lead to higher economic performance.

Indeed, the close link between innovation and entrepreneurship has been fully acknowledged in the last 20 years of entrepreneurship research at the regional level (Fritsch and Storey, 2014) and lies at the heart of the knowledge spillover theory of entrepreneurship (Acs *et al.*, 2009).

The entrepreneurship–innovation nexus is an especially relevant topic in the European context, where a more granular knowledge of where entrepreneurship and innovation potentials can be unlocked is a crucial component for the design and facilitation of smart specialization strategies in EU regions, within the frame of the wider Europe 2020 economic growth strategy, and, in particular, its 'Innovation Union' flagship initiative (Szerb *et al.*, 2013).

In our opinion, studying this issue requires three crucial ingredients:

1   elaborating a more articulated measurement of innovation phenomena in space based on a specific conceptualization and interpretation of the knowledge–innovation logical chain;
2   adopting a more complex measurement of entrepreneurship based on a specific conceptualization and definition of entrepreneurial events; and
3   studying the geography of the interplay and variations of these two phenomena.

In particular, for what concerns innovation, the chapter fully ascribes the view that regions innovate through different modes, according to the presence/absence of some context conditions that allow for the creation and/or the adoption of knowledge and innovation. At the foundations of this view, it is the notion of regional patterns of innovation, defined as a combination of *context conditions* and of *specific modes of performing and linking-up the different phases* of the innovation process (Capello, 2013), where the different phases of the innovation process refer to a logical sequence between knowledge, innovation and economic performance, as in the abstract but consistent 'linear model of innovation'.

For what concerns entrepreneurship, the chapter introduces a subtle interpretation of the notion of entrepreneurship, in line with the function of entrepreneurship identified by Kirzner as a market discovery process (Kirzner, 1997), which stresses the processes of recognition and discovery of underexplored opportunities, the propensity to risk and launch a new business, and the capacity of entrepreneurs to move from discovery to real entrepreneurial action. This interpretation is indeed consistent with much of recent research on entrepreneurship. For example, the knowledge spillover theory of entrepreneurship (Acs *et al.*, 2009) identifies entrepreneurship as the main channel through which new and unexploited knowledge and ideas generated in existing firms (i.e. underexplored opportunities) can be commercialized and brought to the market, thus spurring local growth. Also, the 'National system of entrepreneurship' approach (Acs *et al.*, 2014; Hundt and Sternberg, 2014), which treats entrepreneurship as a systemic phenomenon, recognizes that the entrepreneurial phenomenon has to be understood and studied in the inherent complex, composite and systemic nature deriving from the interaction of individual business behavioural aspects and the institutional context in which entrepreneurship is embedded. In particular, we consider three different but complementary dimensions of entrepreneurship, respectively the potential capacity of opportunities recognition, the orientation to risk and to actual opportunities recognition, and the strategic vision of an entrepreneurial mission. Empirically, the distinction among these three aspects of entrepreneurship has been made possible thanks to a recently released indicator of entrepreneurship, elaborated by DGRegio of the European Commission (Szerb *et al.*, 2013).

The following empirical analysis therefore aims not simply to shed light on the (expected positive) association between entrepreneurship and innovation but also to explore whether their interplay can be more complex than generally assumed in the literature, depending on the specific pattern of innovation and entrepreneurial characteristic considered.

Measuring innovation and entrepreneurship in Europe

Measurement of innovation in this chapter follows a well-defined conceptual approach based on the notion of regional patterns of innovation, defined as a combination of *context conditions* and of *specific modes of performing and linking-up the different phases* of the innovation process (Capello, 2013). The novelty of this approach lies in the fact that it allows to acknowledge and emphasize the heterogeneity of innovation processes in space. In fact, as anticipated in the introduction, the different phases of the innovation process refer to a logical sequence between knowledge and innovation, as in the abstract but consistent 'linear model of innovation' – even if heavily criticized as unrealistic and rooted in the idea of a rational and orderly innovation process (Edgerton, 2004). This framework therefore strongly supports the concept of a 'spatially diversified, phase-linear, multiple-solution model of innovation' (Capello, 2013, p. 137), in which each pattern represents a linearization, or a partial block-linearization, of an innovation process where feedbacks, spatial interconnections and non-linearities play a prominent role.

Importantly, the regional innovation patterns concept stresses complex interplays between phases of the innovation process and the regional context. Context conditions are integral parts of each regional pattern of innovation as, each regional pattern of innovation is a combination of different phases of the linear model of innovation and the presence/absence of regional preconditions allowing for a certain phase to take place.

This approach strongly relies on the conceptual distinction between knowledge and innovation: they are conceptualized as different (and subsequent) logical phases of an innovation process, each phase requiring specific local elements for its development. By consequence, it refuses the generalization of an invention–innovation short-circuit taking place inside individual firms (or territories), as that visible in some advanced sectors, as well as the assumption of an immediate interaction between R&D/higher education facilities on the one hand and innovating firms on the other, because of pure spatial proximity.

Among all possible combinations of innovative processes and context conditions, three main 'archetypes' innovation modes, each of which reflecting a specific piece of literature on knowledge and innovation in space, may be indicated (Capello, 2013):

1   *A science-based pattern*, where highly innovative firms, belonging to high added value and technology-intensive sectors, are expected to cluster

since they seek and require those local conditions, like the presence of universities, research centres and highly advanced human capital, to fully support the creation of knowledge. Moreover, in this group of regions the preconditions to turn knowledge into innovation, like the presence of entrepreneurial spirit and creativity, guarantee the transformation of knowledge into innovation. Given the complex nature of knowledge creation nowadays, this pattern is expected to show a tight interplay among regions in the form of international scientific networks. From the conceptual point of view, this advanced pattern is the one considered by most of the existing literature dealing with knowledge and innovation creation and diffusion (Malecki, 1980, Saxenian, 1994; Audretsch and Feldman, 1996; Mack, 2014; Vogel, 2015).

2   *A creative application pattern,* characterized by the presence of creative (small and medium) enterprises belonging to traditional or medium-tech sectors, curious enough to look for knowledge outside the region – given the scarcity of local knowledge – and creative enough to apply external knowledge to local innovation needs (Foray, 2009; EC, 2010; Licht, 2009). Knowledge providers, supporting local firms' innovative activities, are mostly located outside the region and knowledge exchanges are mainly nourished by cognitive and sectoral proximity (i.e. shared cognitive maps) than by belonging to the same local community.

3   *An imitative innovation pattern,* where firms in traditional sectors, or branches of multinational enterprises, in different sectors, seeking low labour-cost areas to locate their lower value-added functions, base their innovation capacity on imitative processes, that can take place with different degrees of adaptation on already existing innovations. In several cases, regions in this pattern are likely to be characterized by a higher presence of firms with little learning and innovative activities. This pattern is based on the literature dealing with innovation diffusion (Hägerstrand, 1952; Pavlínek, 2002; Varga and Schalk, 2004).

Following this perspective, therefore, an accurate explanation and measurement of regional innovation requires to move beyond traditional input and/or output indicators of innovative activities such as R&D expenditures on GDP or the share of innovative firms in a region.

In a recent work, regional patterns of innovation have been empirically detected by means of a k-means cluster analysis based on a series of indicators capturing the different knowledge and innovation propensity across European regions: namely the regional EU share of total patents, the regional share of firms introducing product and/or process innovation, and the regional share of firms introducing marketing and/or organizational innovation (Capello and Lenzi, 2013).[1] The empirical results show a larger variety of possible innovation patterns than the ones conceptually envisaged, still consistent with the theoretical underpinnings presented before. Two clusters can be associated to our first conceptual pattern, albeit with a clear distinction between basic and applied scientific knowledge; two clusters can be associated to the second pattern, again with some important differences, namely between formal and informal externally sourced knowledge, and one cluster can be associated to the third pattern (Figure 2.1). In particular:

- A *European science-based area (ESBA)*, composed of strong knowledge and innovation producing regions, specialized in basic scientific knowledge and general purpose technologies, with the highest generality and originality of the scientific local knowledge, and the highest degree of knowledge acquisition from other regions. They are mostly located in Germany, with the addition of Vienna, Brussels and Syddanmark in Denmark.

- An *applied science area (ASA)*, similarly made up of strong knowledge-producing regions albeit characterized by a local knowledge base of an applied nature, and by a high degree of knowledge acquisition from other regions. Regions of this type are mostly agglomerated and located in central and northern Europe, namely in Austria, Belgium, Luxembourg, France (i.e. Paris), Germany, Ireland (i.e. Dublin), Denmark, Finland and Sweden, with some notable exceptions to the East such as Prague, Cyprus and Estonia and to South such as Lisbon and Attiki.

- A *smart technological application area (STAA)*, with a high product innovation rate, with a more limited degree of local basic science, but a high level of creativity which enables the translation of external basic and applied science knowledge into innovation with respect to the other four clusters. The knowledge intensity is lower than in the previous two cases, although not negligible. This group of regions includes mostly agglomerated regions in the EU15, such as the northern part of Spain and Madrid, Northern Italy, the French Alpine regions, the Netherlands, Czech Republic, Sweden and the UK.

- A *smart and creative diversification area (SCDA)*, a non-negligible internal innovation capacity, a high degree of local capabilities (i.e. non-scientific and tacit knowledge embedded in professionals), of creativity, and of acquisition of external knowledge embedded in professional capabilities. These regions are mainly located in Mediterranean countries (i.e. most of the Spanish regions, Central Italy, Greece, Portugal), in EU12 agglomerated regions in Slovakia and Slovenia, Poland and the Czech Republic, in a few regions in northern Europe, namely in Finland and the UK.

- An *imitative innovation area (IIA)*, characterized by a low knowledge and innovation intensity but high attractiveness and innovation potentials. Most of these regions are in the EU12, such as all regions in Bulgaria and Hungary, Latvia, Malta, several regions in Poland, Romania and Slovakia, but also in Southern Italy.

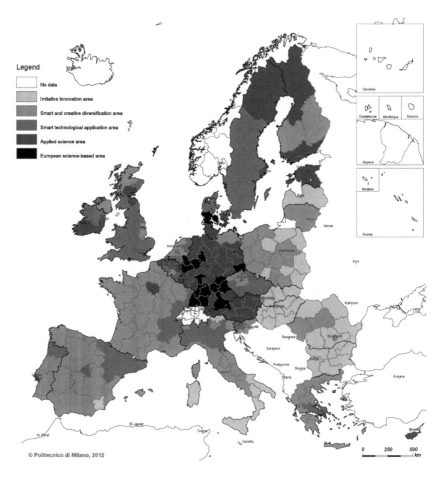

Figure 2.1 Regional patterns of innovation in Europe
Source: Capello and Lenzi, 2013.

Similarly to innovation, measuring entrepreneurship has always been a challenging task. Fritsch and Storey (2014), in their review of three special issues (1984, 1994, 2004) of *Regional Studies* on the subject, state clearly this point although they acknowledge the important advances achieved in the last decade in the quality and availability of entrepreneurship data, in terms of industry, regional and time coverage and disaggregation.

The emphasis put in this chapter on the need of a specific measurement, conceptualization and definition of entrepreneurship has also been facilitated by the existence of a new interesting dataset. The interest of this new database lies precisely in its conception of entrepreneurship measurement; instead of taking the main categories of indicators traditionally used to measure entrepreneurship, namely output, attitude and framework indicators, in this new dataset, called REDI, composite indicators are built for European regions in order to take into account the interplay between individual level desirability and feasibility considerations for entrepreneurial action and the institutional contexts in which these considerations originate.[2] It is exactly this interplay that shapes the final entrepreneurial action and determines the quality and outcomes of this action, as suggested by Acs *et al.* (2014).[3]

The final, aggregate indicator of entrepreneurship, called REDI (i.e. Regional Entrepreneurship and Development Index), is obtained as the simple average of three sub-indexes, called entrepreneurial ability, attitude and aspiration, measuring three main different interesting aspects of entrepreneurship, respectively, potential of opportunities perception, risk orientation and strategic vision. The three main sub-indexes as well are obtained as composite indices representing the outcome of the interplay of 14 pillar variables. In particular, the three main sub-indexes are obtained as arithmetic average of normalized pillars for the respective sub-index multiplied by 100. Also, the 14 pillars as well are composite indicators that merge by interaction up to 76 individual and context (i.e. regional and/or national) level variables, collected over the period 2002–2011.[4]

More in detail, the potential of opportunities perception, defined as the potential capacity to develop start-up activities with high growth potential, is a composite indicator accounting for individuals' interest in self-employment activities. It also considers context characteristics such as the favourability of the business environment, engagement in training, on business sophistication and presence of high-tech manufacturing and knowledge-intensive services in the region. More specifically, it is measured through the entrepreneurial ability composite sub-index available in REDI which combines different dimensions pertaining to entrepreneurial behavioural aspects linked to start-up opportunities, technology adoption, human capital and competition. As anticipated, these characteristics are also measured as composite indicators.

Risk orientation captures risk propensity and actual opportunity recognition, at both individual and regional level, and accounts for the population's self-esteem concerning its ability to start new businesses, its risk acceptance and its capacity to recognize opportunities for new business. Context characteristics included in this sub-index refer to the social status of, and respect for, an entrepreneur, and to the level of corruption and of individual freedom in the local society.

Risk orientation is measured through the entrepreneurial attitude composite sub-index available in REDI integrating different aspects (also obtained as composite indicators) related to risk propensity and opportunity discovery, namely: opportunity perception, start-up skills, risk perception, networking and cultural support. As noted above, also these characteristics are measured as composite indicators.

Strategic vision refers to the distinctive, qualitative, strategy-related nature of entrepreneurial activity and accounts for the capacity of new businesses to develop innovation, to grow, to internationalize and to raise interest in capital markets. Context characteristics considered refer to patent intensity, clustering, connectivity and depth and diversification of capital markets in the region. It is measured through the entrepreneurial aspiration composite index available in REDI merging those characteristics of entrepreneurship associated with (product and process) innovation, capacity for growth, internationalization and globalization, and finance. Similarly to the previous case, these characteristics are measured as composite indicators.

These indices closely fit the conceptualization of entrepreneurship adopted in this chapter, which emphasizes the function of entrepreneurship identified by Kirzner as a market discovery process (Kirzner, 1997), and stresses the processes of recognition and discovery of underexplored opportunities, the propensity to risk and launch a new business, and the capacity of entrepreneurs to move from discovery to real entrepreneurial action.

Table 2.1 provides a more detailed description of the different dimensions and constitutive elements (i.e. pillars in REDI) pertaining to three entrepreneurial sub-indexes.

The next section will discuss the nexus between entrepreneurship and innovation in EU regions: we will do so, first of all, by looking at the geographical distribution of the interplay between innovation and entrepreneurship, and second, by digging into the description of their nexus by applying the above-mentioned conceptualization and measurement of innovation and entrepreneurship.

The innovation–entrepreneurship nexus in European regions

A simplistic definition and measurement of the interplay between innovation and entrepreneurship in space

In order to unravel the geographical nexus between entrepreneurship and innovation, we start by partitioning the European regions on the basis of their innovation intensity and their degree of entrepreneurship. In so doing, we follow a principle of progressive complexification and consider first the aggregate indicator of entrepreneurship and a more traditional measure of innovation, i.e. the share of firms introducing product and/or process innovation in a region.

In particular, we divide regions on the basis of their innovation and entrepreneurship intensity with respect to the EU median value (Figure 2.2).[5] Innovation is measured as the regional share of firms introducing product and/or process innovation; entrepreneurship, instead, is measured through the aggregate REDI index.[6] In each case, four different situations emerge.

Figure 2.3 shows the geographical distributions of these areas. The general impression is one of a positive association between entrepreneurship and R&D as well as between entrepreneurship and innovation. Expectedly, the Eastern and Southern periphery of the Union seems to lag behind in terms of innovation and entrepreneurship; on the other hand, Central and Northern

Europe is better positioned in all respects. Intermediate cases, however, are widespread. In particular, 31 per cent (i.e. 79 regions) do not conform to this trend and show a mismatch between innovation and entrepreneurship. In situations where innovation is greater than the median values but entrepreneurship below the median value, it seems that local entrepreneurs are not able to fully take advantage of the rich innovative environment that exists around them, which is possibly a characteristic of large-firm innovation driven areas. In the opposite case of poor innovative areas, it seems that local entrepreneurs show an extremely dynamic behaviour, that drives them to fully exploit existing local opportunities and to search outside the region for additional ones to feed local needs.

This mixed evidence, possibly, can find some explanation by adopting a more fine-grained measurement (and related founding conceptualization) of both innovation and entrepreneurship. In short, disentangling the possible sources of these unexpected results and better understanding the interplay between innovation and entrepreneurship in space requires acknowledging the relevance of different entrepreneurial characteristics and the existence of different regional innovation patterns.

A complex definition and measurement of the interplay between innovation and entrepreneurship in space

Following again a principle of progressive complexification, Table 2.2 first introduces a more complex definition of innovation, i.e. the one based on the regional innovation patterns concept, and partitions European regions according to their regional pattern of innovation and the classification of the innovation–entrepreneurship nexus introduced above. Results are rather interesting as they signal a remarkable concordance between the two classifications. In particular, all regions in the imitative group are also poor entrepreneurial-innovation regions. On the other hand, regions in the most innovative group (i.e. the European science-based area) are almost all in the most innovative and entrepreneurial case. Intermediate cases are possibly of greater interest. Regions in the applied science area figure prominently in the most innovative and entrepreneurial case, but also not negligibly in the group of large innovative firms, similarly to regions in the smart technological application area, which, however, are also well represented in the poor innovative but entrepreneurial case. Differently, the smart and creative diversification area includes the largest group of regions in the poor innovative but entrepreneurial case even if many regions of this pattern are in the least innovative and entrepreneurial group. In short, these results suggest that the science-based patterns are characterized by high innovation, in most of the cases associated to high entrepreneurship. On the other hand, the application based patterns are characterized by the relative greatest entrepreneurship, even if not always complemented by high innovation. Finally, the imitative group is weak in both respects.

Table 2.1 Measurement of entrepreneurial characteristics

	Measurement
Entrepreneurship	**REDI**
Potential of opportunities recognition	**Entrepreneurial ability**
Opportunity start-up	Depends on the opportunity motivation of the population and on the favourability of the business environment
Technology adoption	Depends on the share of new/nascent businesses in high-tech sectors, on technological readiness and on employment in knowledge intensive and high-tech firms
Human capital	Depends on entrepreneurs educational attainment (i.e. secondary school) and on the region's population engagement in training and life-long learning
Competition	Depends on the number of competitors, on the nature of competitive advantage and on business sophistication
Risk orientation	**Entrepreneurial attitude**
Opportunity perception	Depends on population's capacity to recognize opportunities and on market agglomeration, which reflects the size of the market, population growth, urbanization and accessibility in a region
Start-up skill	Depends on population's self-esteem about its ability to start (successfully) a new business and on the quality of the education in the region
Risk perception	Depends on population's risk acceptance and on the general business risk proxied by the country business disclosure rate
Networking	Depends on a population's knowledge of entrepreneurs and on the technological readiness of a region
Cultural support	Depends on population's views about the carrier possibilities and the social status and respect of entrepreneurs, on individual freedom and on the level of corruption

Strategic vision

Product innovation — Depends on the capacity to create new products and on the region's potential to patent and create scientific publications

Process innovation — Depends on the capacity to create new processes and to invest in R&D

High growth — Depends on the presence of high-growth firms and on the presence of clusters in a region

Globalization — Depends on firms' export potential and on the connectivity of a regions

Financing — Depends on the informal financing provided by friends, relatives or business angels, on the access to different capital sources and on depth of capital markets

Entrepreneurial aspiration

Source: Adapted from Szerb *et al.* (2013). For additional details on the computation of REDI's constitutive sub-indices and related sub-dimensions (i.e. pillars) see: http://ec.europa.eu/regional_policy/sources/docgener/studies/pdf/regional_entrepreneurship_development_index.pdf.

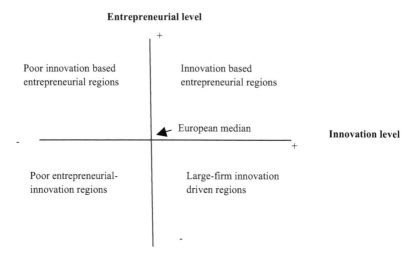

Figure 2.2 A classification of regions according to their innovation–entrepreneurship nexus

In order to dig further into this intriguing but somewhat unexpected result, Table 2.3 displays the results of an ANOVA exercise implemented on the entrepreneurial variables and their constitutive pillars, presented in the previous section, across the five patterns of innovation and in Europe.

The first impression one can get from Table 2.3 is similar to the results of Figure 2.3 and Table 2.2, that of a general positive association between entrepreneurship and innovation as more advanced innovation patterns show higher values of entrepreneurship aggregate index (REDI) as well as of its sub-indices (i.e. potential of opportunities recognition, risk orientation and strategic vision). Interestingly enough, the values reported by the smart technological application area are not statistically different from those of the two science-based patterns (i.e. applied science area and European science-based area). However, these three groups of regions show significantly higher sub-indices values than the smart and creative diversification area (which in turn shows greater values than the imitative innovation area). Therefore, patterns that pertain to the same conceptual archetype innovation model can show distinctive entrepreneurial traits.

The positive association between entrepreneurship and innovation is confirmed by looking at the values of the constitutive pillars of the potential of opportunities recognition across the five regional patterns of innovation. The divide between the imitative innovation area on the one hand, the smart creative diversification area on the other (both showing most of the times values lower than the European average), and the other three groups of regions emerges once more, suggesting that favourable potential conditions

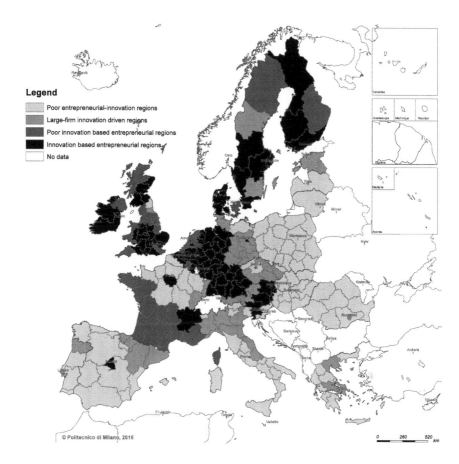

Figure 2.3 A classification of regions according to their innovation–entrepreneurship nexus

for entrepreneurship are actually associated with enhanced innovation and more advanced innovation models.

A similar conclusion can be drawn by looking at the values of the constitutive pillars of risk orientation across the five regional patterns of innovation. In this case, however, an exception stands out. In fact, the risk perception pillar, which accounts for the propensity to take the risk of launching a new (possibly innovative) business, has similar values in the imitative innovation area and in the European science-based area but lower in these two groups with respect to the others. It is not possible to exclude ex ante that this different distribution of entrepreneurial characteristics values across patterns can be explained by the fact that entrepreneurship in the smart and creative diversification area is more driven by necessity due to limited job opportunities and sluggish labour markets than by actual opportunities to enter successfully

Table 2.2 The interplay between regional patterns of innovation and the innovation–entrepreneurship nexus

	Imitative innovation area	Smart and creative diversification area	Smart technological application area	Applied science area	European science-based area	Total
Poor entrepreneurial-innovation regions	100%	65%	3%	-	-	88
Large-firm innovation driven regions	-	6%	28%	31%	10%	41
Poor innovation-based entrepreneurial regions	-	28%	16%	6%	-	38
Innovation-based entrepreneurial regions	-	1%	52%	63%	90%	85
Total	30	86	67	49	20	252

Note: percentage values are computed on column totals.

in the market. Still, risk orientation can be an important leverage to initiate innovative entrepreneurial projects, introducing novelties and adjusting and replicating external knowledge-search behaviours in the absence of objective entrepreneurial opportunities (Acs *et al.*, 2014).

Importantly, a strategic vision of entrepreneurship turns to be quite important also in patterns that are not science-based. In particular, entrepreneurship oriented to innovation seems particularly predominant in contexts in which innovation is based on the creative application of external knowledge. This case can be explained by the high creative and strategic entrepreneurship existing locally that allows to exploit external knowledge for local needs. In these innovation patterns, product innovation does not differ significantly in the application patterns from the science-based patterns and process innovation is even higher in the application patterns than in the European science-based area. For what concerns the other constitutive pillars of this entrepreneurial sub-index, instead, the usual divide between the imitative innovation area on the one hand, the smart creative diversification area on the other, and the other three groups of regions is confirmed once again.

All in all, results confirm the positive association between entrepreneurship and innovation but also highlight important spatial variations according to the different patterns of innovation considered and the entrepreneurial dimension considered. In particular, risk orientation as well as a strategic vision of entrepreneurship based on product and, mostly, process innovation are especially relevant in innovation patterns that are not science-based and could be therefore thought of as important leverages to act on in order to stimulate innovation and, in turn, regional performance.

Conclusions

The entrepreneurship–regional growth nexus has been subject to a large number of conceptual and empirical studies, leaving still open a definitive conclusion. This chapter has proposed to take a step back better to understand and reconcile this growing but often incomparable, conflicting and, sometimes, 'stand-alone' pieces of evidence. In particular, it has proposed to look once more at the geography of the interplay between innovation and entrepreneurship rather than looking directly at the relationship between entrepreneurship and regional performance, so to underline in which situations entrepreneurship is actually accompanied with innovation and can therefore lead to higher economic performance. In so doing, it has adopted a specific conceptualization and measurement of entrepreneurship as well as a specific framework and measurement to interpret the spatial variants of innovation processes across regions. The interplay of these dimensions has turned out also in empirical terms an interesting way to acknowledge some situations, which do not strictly conform to the standard positive association between entrepreneurship and innovation proposed in the literature.

Our results in fact confirm the positive association between entrepreneurship and innovation, indeed, more advanced innovation patterns are characterized by increasing levels of entrepreneurial characteristics. Our results however also show that in a context relatively weaker in terms of knowledge creation (i.e. in creative application patterns in the jargon of this chapter), entrepreneurs can be able to search for the right knowledge outside the region and turn it into product and mostly process innovation, a situation not yet empirically underlined. Similarly, risk orientation can be an important driver spurring curiosity for novelty and the launch of new innovative projects. The descriptive results also suggest that the cause–effect chain between entrepreneurship and regional growth is highly mediated by the innovative environment (i.e. pattern) in which entrepreneurs are embedded, as recent empirical analyses have already pointed out (Capello and Lenzi, 2015).

All these results underline that policy interventions on entrepreneurship have to be tailored upon the innovative context conditions and the type of entrepreneurship already present in a region, in line with the current debate on smart specialization which promotes supporting, strengthening and diversifying the virtuous aspects of regional specificities, and, particularly in this case, of each regional innovation process (Boschma, 2014; Camagni *et al.*, 2014; Coffano and Foray, 2014).

Notes

1 For further details on the variables used in the cluster analysis and the variables representing the key regional distinctive traits of the different groups of regions see Capello and Lenzi (2013).

2 Because of data availability constraints, REDI and its constitutive pillars and sub-indicators have been developed at a mix of NUTS1 and NUTS2 level, depending on the country and for all EU-27 countries with the exception of Bulgaria, Cyprus, Luxembourg and Malta. For those countries in which data were available at NUTS1 level only, data at NUTS2 level have been extrapolated by assigning the same value to all NUTS2 regions belonging to the same NUTS1.

3 Acs *et al.* (2014) highly criticized the traditional indicators of entrepreneurship; it is their opinion that despite their own merits, traditional indicators fail to take into account the context in which new firms come to operate and the process through which new business come to operate (output indicators), the feasibility and actual realization of entrepreneurial events (attitude indicators), the process through which new business come to operate (framework indicators).

4 For more details on the operazionalization, computation and the rational of the choice of the variables used for obtaining the composite sub-indexes (i.e. ability, attitude and aspiration) and their respective pillars see Szerb *et al.* (2013).

5 The median value has been preferred to the average value because of the skewed distribution of the variables, especially the innovation one.

6 Innovation data have been compiled by the authors starting from EUROSTAT CIS data at the national level. In particular, innovation data are based on national CIS 2006 wave figures (covering the 2004–2006 period) next estimated at the NUTS2 level. For additional information on innovation data see Capello and Lenzi (2013).

Table 2.3 Entrepreneurial characteristics across regional patterns of innovation

	Imitative innovation area	Smart and creative diversification area	Smart technological application area	Applied science area	European science-based area	EU average	p-value
Entrepreneurship	26.66	41.24	54.27	55.54	57.77	47.06	p<0.01
Potential of opportunities recognition	19.74	38.92	56.52	56.98	60.53	46.54	p<0.01
Opportunity start-up	1.43	3.25	5.12	4.98	5.08	4.01	p<0.01
Technology adoption	10.17	20.02	26.28	27.20	30.51	22.74	p<0.01
Human capital	8.62	13.85	19.39	20.96	21.68	16.71	p<0.01
Competition	11.69	21.24	29.89	29.38	31.56	24.81	p<0.01
Risk orientation	27.92	40.35	55.83	52.57	54.07	46.45	p<0.01
Opportunity perception	3.52	4.94	8.01	7.05	8.44	6.27	p<0.01
Start-up skill	56.50	75.92	99.70	114.68	105.37	89.80	p<0.01
Risk perception	2.86	3.41	4.69	3.51	2.98	3.67	p<0.01
Networking	6.67	11.10	16.29	16.62	16.89	13.48	p<0.01
Cultural support	1.25	2.63	3.74	3.34	3.70	2.98	p<0.01
Strategic vision	32.29	44.45	50.47	57.07	58.70	48.19	p<0.01
Product innovation	2.96	4.51	4.52	4.89	5.26	4.46	p<0.01
Process innovation	0.27	0.36	0.35	0.36	0.28	0.34	p<0.01
High growth	0.64	0.57	0.85	0.85	0.92	0.074	p<0.01
Globalization	2.06	2.33	2.46	3.13	3.24	2.56	p<0.01
Financing	55705	100410	105520	196278	169972	120608	p<0.01

Note: The table reports non-normalized values of the pillar variables across patterns. This explains their different range of variations.

References

Acs, Z.J., Autio, E. and Szerb, L. 2014. National systems of entrepreneurship: measurement issues and policy implications. *Research Policy*, 43: 476–494.

Acs, Z.J., Braunerhjelm, P., Audretsch, D.B. and Carlsson, B. 2009. The knowledge spillover theory of entrepreneurship. *Small Business Economics*, 32(1): 15–30.

Audretsch, D.B. and Feldman, M.P. 1996. R&D spillovers and the geography of innovation and production. *American Economic Review*, 86(3): 630–640.

Boschma, R. 2014. Constructing regional advantage and smart specialization: Comparison of two European policy concepts. *Scienze Regionali – Italian Journal of Regional Science*, 13(1): 51–68.

Camagni, R., Capello, R. and Lenzi, C. 2014. A territorial taxonomy of innovative regions and the European regional policy reform: Smart innovation policies. *Scienze Regionali – Italian Journal of Regional Science*, 13(1): 69–106.

Capello, R. 2013. Territorial patterns of innovation. In Capello, R. and Lenzi, C. (eds) *Territorial Patterns of Innovation: An Inquiry on the Knowledge Economy in European Regions*, pp. 129–150. Oxford: Routledge.

Capello, R. and Lenzi, C. 2013. Territorial patterns of innovation in Europe: A taxonomy of innovative regions. *Annals of Regional Science*, 51(1): 119–154.

Capello, R. and Lenzi, C. 2015. The entrepreneurship and regional growth nexus: The role of regional innovation modes and entrepreneurial behavioural characteristics. *Small Business Economics*, forthcoming.

Coffano, M. and Foray, D. 2014. The centrality of entrepreneurial discovery in building and implementing a Smart Specialization Strategy. *Scienze Regionali – Italian Journal of Regional Science*, 13(1): 33–50.

EC, Commission of the European Communities. 2010. *Regional Policy Contributing to Smart Growth in Europe*, COM(2010)553, Brussels.

Edgerton, D. 2004. The linear model did not exist: Reflections on the history and historiography of science and research in industry in the twentieth century. In Grandin, K., Worms, N. and Widmalm, S. (eds) *The Science Industry Nexus: Science History Publications*, pp. 31–57. Sagamore Beach: Science History Publications.

Foray, D. 2009. Understanding smart specialisation. In Pontikakis, D., Kyriakou, D. and van Bavel, R. (eds) *The Question of R&D Specialisation*, pp. 19–28. JRC, European Commission, Directoral General for Research, Brussels.

Fritsch, M. and Storey, D.J. 2014. Entrepreneurship in a regional context: Historical roots, recent developments and future challenges. *Regional Studies*, 48(6): 939–954.

Hägerstrand, T. 1952. The propagation of innovation waves. *Lund Studies in Geography, Human Geography*, 4: 3–19

Hundt, C. and Sternberg, R. 2014. Explaining new firm creation in Europe from a spatial and time perspective: A multilevel analysis based upon data of individuals, regions and countries. *Papers in Regional Science*, DOI: 10.1111/pirs.12133.

Kirzner, I. M. 1973. *Competition and Entrepreneurship*. Chicago: University of Chicago Press.

Kirzner, I. M. 1997. Entrepreneurship discovery and the competitive market process: An Austrian approach. *Journal of Economic Literature*, 35(1): 60–85.

Licht, G. 2009. How to better diffuse technologies in Europe. *Knowledge Economy Policy Brief* 7: 1–5.

Mack, E. 2014. Broadband and knowledge intensive firm clusters: Essential link or auxiliary connection? *Papers in Regional Science*, 93(1): 3–29.

Malecki, E. 1980. Corporate organisation of R&D and the location of technological activities. *Regional Studies*, 14(3): 219–234.

Pavlínek, P. 2002. Transformation of central and east European passenger car industry: Selective peripheral integration through foreign direct investment. *Environment and Planning A*, 34: 1685–1709.

Saxenian, A.L. 1994. *Regional Advantage: Culture and Competition in Silicon Valley and Route 128*. Boston: Harvard University Press.

Schumpeter, J.A. 1934. *The Theory of Economic Development: An Inquiry into Profits, Capital, Credit, Interest and the Business Cycle*. Translated from the German by Redvers Opie. New Brunswick and London: Transaction Publishers.

Szerb, L., Acs, Z.J., Autio, E., Ortega-Argilés, R. and Komlosi, E. 2013. REDI: The regional entrepreneurship and development index. Available at: http://ec.europa.eu/regional_policy/sources/docgener/studies/pdf/regional_entrepreneurship_development_index.pdf.

Varga, A. and Schalk, H. 2004. Knowledge spillovers, agglomeration and macroeconomic growth: An empirical approach. *Regional Studies*, 38(8): 977–989.

Vogel, J. 2015. The two faces of R&D and human capital: Evidence from Western European regions. *Papers in Regional Science*, 94(3): 525–551.

3 Networking practices and networking cultures

Ben Spigel

Introduction

The size and structure of entrepreneurs' social networks are key predictors of survival and success. Entrepreneurs with larger social networks with diverse arrays of contacts tend to have better access to the financial and knowledge resources they need to discover and exploit opportunities in the marketplace (Arenius & de Clercq, 2005). There is significant evidence that the size and quality of entrepreneurial networks differs between regions and nations, which is both a cause and consequence of the continued clustering of entrepreneurial activity (Rutten, Westlund, & Boekema, 2010). The heterogeneous geography of entrepreneurial networks has far-reaching consequences for regional economic development and innovation: regions where entrepreneurs have smaller or less resource-rich networks will also tend to have lower rates of firm formation and growth.

But despite the voluminous research on entrepreneurial networks there are two significant research gaps. First, networking is frequently viewed as an instrumental activity entrepreneurs engage in to support their firms: entrepreneurs are expected to access their social capital in the same way they might access their financial capital in a bank. This ignores the fact that *social* networks are deeply *social* things, intertwined in personal life. Second, with exceptions such as Klyver and Foley (2012) and Lang, Fink and Kibler (2014), there is little discussion about the relationship between regional cultural and economic structures with the networking practices entrepreneurs employ. As a result, while there is substantial evidence of variation in networking patterns between regions, we know comparatively little about the processes linking cultural outlooks and networking practices.

This chapter addresses these issues through an investigation of the networking practices of technology entrepreneurs in two Canadian cities: Waterloo, Ontario and Calgary, Alberta. The chapter argues that studying entrepreneurs' use of networks and social capital requires us to understand entrepreneurs' attitudes toward the act of entrepreneurship itself. These attitudes are constructed within larger cultural milieu about the purpose and role of entrepreneurship. The chapter employs a Bourdieuian perspective in which these practices are the outcomes of entrepreneurs' individual habitus and the unwritten "rules" of a local field. The next section discusses previous research on the role of networks within the entrepreneurship process and previous thought about how regional and national cultures influence how entrepreneurs network. This is followed by a discussion of Bourdieu's sociology of practice in the context of the entrepreneurship process. The third section introduces the two case studies and the qualitative methodologies used to study networking within them. This is followed by a comparison of networking practices between technology entrepreneurs in Calgary and Waterloo. The chapter concludes by arguing for a more nuanced perspective of entrepreneurial networks.

The geography of entrepreneurial networks

The role of networks in venture creation and survival

Entrepreneurship is a relational and social process (Dodd & Anderson, 2007). Knowledge about new opportunities, technologies, or changes in the marketplace gained through trusted social contacts is the foundation of entrepreneurial innovation (Stuart & Sorenson, 2003). Entrepreneurs with larger and more diverse networks have been shown to have superior access to financing (Shane & Cable, 2002), novel ideas (Powell, White, Koput, & Owen-Smith, 2005), and are better able to capitalize on opportunities in the marketplace (Anderson & Miller, 2003). There is little doubt that the size and structure of entrepreneurs networks significantly influence every stage of the entrepreneurship process. Since the introduction of Aldrich and Zimmer's (1986) "network success hypothesis," there has been a sustained research interest in how entrepreneurs employ their social networks (see Hoang & Antoncic, 2003). The structure and content of an entrepreneur's network affect both their original intentions to start a firm and their ability to identify potential opportunities (Burt, 2004). As start-ups grow, entrepreneurs draw on knowledge from their networks to find new clients and resources (Lechner & Dowling, 2003). Later in their growth processes, entrepreneurs employ their social networks to find investment from angel investors and venture capitalists (Sorenson & Stuart, 2001).

The structure of a network ultimately determines its usefulness to entrepreneurs. Networks with a diverse array of actors in them will contain more unique sources of useful information to help the entrepreneur identify new opportunities (Anderson & Jack, 2002). Entrepreneurs must balance the need for a small number of very strong ties that are based on long-term trust which can provide substantial resources such as investments with a larger number of weaker ties with unique resources at their disposal (Granovetter, 1973; Jack, 2005). But there is little evidence for an optimal structure of entrepreneurial networks. While Brüderl and Preisendöfer (1998) find that higher proportions of strong ties help firm survival due to the increased support and resources entrepreneurs can draw from them, Uzzi (1996) argues that a surfeit of strong ties leads to over-embeddedness, where entrepreneurs are prevented from taking the most economically rational actions, such as finding a new supplier, due to their long-standing ties of trust.

Networking practices vary across time and place, with particular networking practices becoming part of regional or organizational routines. This includes factors such as the proclivity of workers to meet up and share knowledge over drinks, freely share advice and contacts, or to protect propriety knowledge by avoiding contact with outside groups (Henry & Pinch, 2001; James, 2005). These locally specific networking practices evolve over time in response to both regional cultural outlooks toward networking and the organizational practices of dominant companies in the area. Saxenian (1994) highlights how the open organizational cultures of Silicon Valley firms like Hewlett-Packard led to a networking culture there as opposed to how Boston's closed corporate culture discouraged networking and knowledge-sharing amongst technology firms.

Within the entrepreneurship literature networking is frequently treated as an instrumental act rather than as a part of the entrepreneur's social life. A very large stream of the entrepreneurial networking literature ignores the heterogeneity of networking practices and instead adopts a universalist model of entrepreneurial social capital (Klyver, Hindle, & Meyer, 2008). From this perspective, networking is a strategic activity entrepreneurs engage in to increase the amount of resources they can draw upon in their networks. This is not to say that networking is unimportant in the entrepreneurship process: Neff, Wissinger and Zukin (2005) demonstrate the overriding importance of network building in technology entrepreneurship. Rather, there is a tendency to view networking as a purely purposeful strategy rather than as intrinsically social activity. Networks are simplified into their structural characteristics such as size and diversity, eliminating subtle differences in networking practices and content.

Bourdieuian perspectives on entrepreneurial networking

A Bourdieuian perspective is a useful way to frame the interactions between regional culture and networking practices while maintaining a socially situated view of networking itself. A Bourdieuian analysis view entrepreneurs' practices – the day-to-day actions they carry out in pursuit of their goals – as the emergent objective rules and power hierarchies of a social system (*field*) as well as how these rules are understood, interpreted, and internalized by the individual within their *habitus* (Bourdieu, 1977; Bourdieu & Wacquant, 1992; Swartz, 1997). The explicit focus on the interplay between individual outlooks and structural forces helps resolve the tension between deterministic structuralism and contextless methodological individualism. In a Bourdieuian perspective on entrepreneurship, fields represent the beliefs and outlooks regarding the social and economic role of entrepreneurship (Karataş-Özkan, 2011; Spigel, 2013). These can be understood as the "rules of the game" of entrepreneurship (Stringfellow, Shaw, & Maclean, 2014). This includes attributes like the social status accorded to entrepreneurs (which differs based on their industry and perceived "innovativeness"), the acceptability of the risk and extra work entrepreneurs take on, and the importance of entrepreneurship as a personal vocation instead of a way to make money. Fields also determine the value of different kinds of capital such as economic capital (money or profits), social capital (social connections and resources), or cultural capital (the social status of building a successful business). As Bourdieu (1986) argues, while the overriding goal of modern capitalism might be the accumulation of economic capital other types of capital may be more or less important in particular fields. Actors structure their practices around acquiring the types of capital they believe are most valuable.

However, fields are not the cause of entrepreneurial practices. Actors understand the rules of the field through their habitus, their internalized dispositions toward entrepreneurial actions (Bourdieu, 1990, 2005). The habitus is developed through an individual's experiences within fields, particularly through their education. This means that the seemingly objective rules of a field are interpreted differently by actors based on their habitus, allowing them to develop new types of practices. These practices are not rote responses to a standard field of expectations but rather a form of improvisation and experimentation based on individual goals, skills, and situations. Practices therefore emerge out of the combination of the rules of a field and the diverse ways in which actors interpret those rules through their habitus and form their own goals based on what forms of capital they think are most valuable to them. There is no one singular "entrepreneurial habitus." Entrepreneurs develop unique outlooks and orientations toward the rules and structures of the local field through their previous educational and work experience which in turn enables or discourages certain types of practices.

We can speak of two major fields that influence the technology entrepreneurship process: the field of technology entrepreneurship (FTE) and the local field. The FTE refers to the norms, expectations, and understandings about starting and running a high-tech firm. This includes both the expectation to take on substantial personal risk and to work far beyond the normal 40-hour work week but also social expectations of what an entrepreneur should look and act like (Centner, 2008). For instance, technology entrepreneurs must adopt specific forms of dress and presentation (e.g. not wearing a suit on a daily basis) but at the same time are often expected to violate the "rules" of the field in order to demonstrate their innovativeness and independence (de Clercq & Voronov, 2009a, 2009b). Meeting these expectations help entrepreneurs build the symbolic capital necessary to be seen as legitimate entrepreneurs (de Clercq & Voronov, 2009c).

The local field represents the informal social rules regarding entrepreneurship that have developed within the region. Local fields develop over time based on the region's economic and social history, the influence of

prominent local "success stories" or dominant employers, and the influence of outside forces such as national policies and global capitalism
(Fligstein & McAdam, 2012). The local field is of particular importance
because entrepreneurial actors are continually embedded in it, making its
influence hard to ignore (Spigel, 2013). Entrepreneurs depend on resources
contained within the local field and therefore must meet the expectations
of entrepreneurship within the field – if they do not it will be difficult to
be taken as a legitimate economic actor deserving of resources. But beyond
strategically hewing to the rules of the local field, entrepreneurs internalize
these rules on a non-conscious level, normalizing particular actions or outlooks to the point where they become part of actor's unspoken repository
of practices. In this sense, the local field represents the dominant norms,
rules, and outlooks of the regional culture.

Entrepreneurs operate within the context of both the FTE and their local
field and must develop practices that balance the often conflicting demands
of these two fields if they are to be viewed as legitimate entrepreneurs. This
contributes to the formation of distinct patterns of regional entrepreneurial
practices and identities, such as the ways in which technology entrepreneurs in Montana balanced the "Silicon Valley" entrepreneurial outlooks
of their industry against the risk averse cultural orientation of their rural
community (Gill & Larson, 2014) or how entrepreneurs in Salt Lake City
sought compromises between the expectations of the FTE for long work
hours with the preference of the local Mormon community to spend that
time with family and friends (James, 2005). Entrepreneurs strike these bal-
ances based on their own internalized views of what practices make sense
given their field-based contexts. Differences in entrepreneurial practices and
processes between regions are therefore not due to the deterministic force of
social structures but are rather the outcome of individual actors choosing
what practices they think makes sense given their habitus-informed understanding of the rules of the fields they operate in along with their goals and
visions for the future.

From this perspective, entrepreneurial networking is not simply an economic activity. Networking – the act of purposefully connecting with others
to gain resources, information, and support – is part of an entrepreneur's
larger social life. The extent to which entrepreneurs choose to dedicate time
and energy toward networking (such as by attending networking events, meeting colleagues after work, or helping to organize entrepreneurial events) as
opposed to developing the business or spending time with their family depends
on how the social and symbolic capital produced by networking is valued
against other forms of capital. Entrepreneurs' networks provide two main
types of benefits: the actual resources entrepreneurs can gather from them
(social capital) and the legitimacy of being associated with other dynamic,
growth-oriented entrepreneurs (symbolic capital). While this symbolic capital
is prized within the FTE and is an important way of attracting investment, it
may be less important in some local fields.

Table 3.1 Demographic and economic data for Calgary and Waterloo

	Calgary	Kitchener-Waterloo	Canada
Population	1,096,833	477,160	33,476,688
Self-employment rate (%)	11%	9%	11%
High-tech firm formation per 100,000 residents (2001–2006)	12.8	7.5	6.1
Labor force in natural and applied science occupations (%)	11.9%	8.9%	7.2%
Population with bachelor's degree or higher (%)	28.8%	21.6%	20.9%
Bachelor's degrees or higher in STEM fields (%)	15.1%	11.6%	9.8%
GDP per capita (2007 dollars)	$73,151	$50,161	$45,704
Average size of VC investment, 2000–2011 (2007 dollars)	$2,866,391	$1,979,297	$239,583
VC investments per 100,000 residents (2000–2011)	17.9	19.5	17.9

Sources: Statistics Canada (2011); Conference Board of Canada (2012); Thomson Reuters (2013).

Entrepreneurial networking in Calgary and Waterloo, Canada

Case study selection and methods

Two case studies of technology entrepreneurs in Calgary and Waterloo, Canada are used to explore the socially situated nature of entrepreneurial networking within local fields. While both cities are centers of Canadian technology entrepreneurship and innovation, entrepreneurs exhibited very different networking practices. These differences appear related to the structure of the cities' local fields and how the rules and norms of the FTE are interpreted through them. As shown in Table 3.1, both regions have high rates of high-tech entrepreneurship, highly educated labor forces and active entrepreneurial investment environments.

Semi-structured interviews were conducted with technology entrepreneurs between 2010 and 2011. A pool of entrepreneurs in six high-tech sectors[1] who founded companies after 1990 was constructed using Scotts Corporate Directory, a Canadian business directory. Entrepreneurs were contacted at random from this list in order to avoid a bias toward more prominent start-ups. The response rates in Calgary and Waterloo were 27 percent and 35 percent, respectively. Along with questions about firm history, the regional economic environment, and networking practices, firm founders were asked to identify which of six entrepreneurial resources they had drawn on while building their firm. These resources where: (1) an emergency loan of less than $5,000; (2) an investment greater than $10,000; (3) a loan provided at a reduced interest rate; (4) a referral to a skilled accountant or lawyer; (5) a referral to someone

Table 3.2 Type and location of interviews

	Calgary	Waterloo	Total
Entrepreneurs	28	23	51
Investors	5	5	10
Economic development officials	6	4	10
Total	39	32	71

who could aid the entrepreneur with marketing or sales; and (6) a referral to an employee they have hired. This provides a proxy for their overall levels of social capital. To triangulate findings within entrepreneurs additional interviews were conducted with selected economic development officials and investors (see Table 3.2).

Networking for business development in Calgary

Calgary's entrepreneurial economy is propelled by a boom in the province's oil and gas sector. While oil and natural gas have been extracted in Alberta since the early twentieth century, the development of the Athabasca Oil Sands in northern Alberta has helped Calgary develop into a command and control center for Canada's energy industry (Chastko, 2004). The oil and natural gas industry dominates the regional economy, accounting for more than 20 percent of the region's overall GDP (Conference Board of Canada, 2012). The industry's size gives it considerable influence over Calgary's local field. The organizational structures and norms of foreign energy firms were imprinted in the region through the pervasive economic presence of US subsidiaries, which served as a major training ground for most of the industry's engineers, managers, and technicians. During the 1950s and 1960s, many executives of domestically owned oil firms in Calgary had spent time in large US-controlled firms like Imperial Oil (the largest oil firm in Canada) and absorbed these firms' organizational culture (Foster, 1979). This trend is still visible. Many interviewees initially moved to Calgary to take a job at one of the larger oil firms. They internalized the industry's unwritten rules working at these firms and carried the outlooks with them when founded their own start-ups. Much of the region's technology entrepreneurship is oriented toward serving this industry. In the words of a local angel investor: "a lot of technology [entrepreneurship] happens here. I think that 10 or 11 percent of the population is employed in the technology sector, but 99.5 percent of that is in technology for oil and gas" (C135). The region's other major technology cluster of wireless communications developed out of the pressing need for resilient communication links between oil wells in the north of Alberta (Lucas, Sands, & Wolfe, 2009).

The influence of the oil and gas industry has led to the development of several significant elements within the local field that have affected the nature of entrepreneurial networking within it. The high value placed on acquiring economic capital and the lower value of the symbolic capital produced by successful entrepreneurship is the most important aspect of the local field. Interviews with entrepreneurs, investors, and economic development officials suggest that actors are motivated by the goal of accumulating economic capital (the profits from an entrepreneurial endeavor or investment) rather than the symbolic capital (the social prestige) of starting and running an advanced technology company. As an early sociological study of Calgary's oil industry argued: "the spirit of competition [in Calgary] translates into the need to make a good living and achieve career success" (House, 1980, p. 80).

The focus on economic capital appears to have reduced the symbolic capital associated with entrepreneurship within the local field. While entrepreneurship has the potential to generate a great deal of economic capital, an entrepreneurial windfall takes years of preparation. During this time, entrepreneurs typically makes less money while working far harder and bearing far more risks than regular employees. In some fields, the symbolic capital and prestige of being an entrepreneur can compensate for the lower economic returns, but this does not appear to be the case in Calgary. The economic capital produced by entrepreneurship, rather than the act of entrepreneurship itself, is most valued within Calgary's local field. While there is a burgeoning technology entrepreneurship scene unrelated to the oil and gas industry, it is relatively small compared to this much larger market. This means that the field of technology entrepreneurship has less influence within the local field than the field of the oil and gas industry.

One outcome of this is the reduced importance of networking with other entrepreneurs to build symbolic capital or develop new skills. In situations where there is little symbolic reward for being an entrepreneur, there is little incentive to spend what little available time an entrepreneur has on spending time with other entrepreneurs. Calgarian entrepreneurs had lower levels of social capital than their counterparts in Waterloo. Of the six social resources entrepreneurs were asked about, Calgarian entrepreneurs utilized an average of 2.04 compared to 2.61 in Waterloo. Furthermore, a lower proportion of entrepreneurs in Calgary said they discussed entrepreneurship with their friends and family. Only 46 percent of entrepreneurs in Calgary reported this kind of conversation, compared with 70 percent in Waterloo. This indicates that entrepreneurs in Calgary are less likely to have discussed entrepreneurial challenges with others and that they gather fewer entrepreneurial resources from their social networks.

Despite their low rates of networking with other entrepreneurs there are several reasons to suggest that Calgarian entrepreneurs are otherwise very active networkers. The petroleum industry is largely project based and oil exploration and development requires the cooperation of many different firms. This means that firms must constantly be looking for new projects and contracts; knowledge of which is spread through social networks rather than formal requests for proposals according to interviewees. As one entrepreneur suggested: "oil and gas is an old-boys network. It's a lot more the face-to-face handshaking than a 'let's pulverize you'" (C129). Calgary is filled with constant buzz and knowledge exchange as firms seek out new partners, clients, and suppliers for temporary projects. Second, there appears to be a great deal of worker mobility. Employees frequently move between firms, bringing their networks of contacts with them. As one entrepreneur described:

> we've got 27 people or so on staff and they've come from other firms that have contacts from other firms, and when they start working here they might call their contacts up and say I'm not longer at [old firm] and I'm over at [new firm] now and let's go for lunch.
>
> (C114)

Calgary is marked by high levels of networking between entrepreneurs and clients but relatively lower levels of networking between entrepreneurs themselves. Networking is understood as a way to create and sustain business relationships rather than a way to build entrepreneurial skills and solve technical and business problems. This is evidenced by entrepreneurs' resistance to participating in public entrepreneurship training and networking programs. While Innovate Calgary (a publicly funded start-up incubation facility) runs entrepreneurship training and networking programs, none of the entrepreneurs interviewed – even Innovate Calgary's own tenants – reported participating in them. Nor did entrepreneurs attend other networking events hosted by the Chamber of Commerce or other local organizations because they did not believe attending these programs would lead to increased profits. As one entrepreneur explained:

> there's been quite a few different entrepreneurship groups, but what I have found is that most of them are there because they think they're going to get a chance to meet potential clients. What it ends up being is a bunch of people like themselves.
>
> (C104)

Networking practices in Calgary have developed within the context of the region's local field with little influence from the outside FTE. The purpose of networking is to develop business relationships with local clients in order to produce economic capital rather than spending time socializing with other entrepreneurs in order to develop both their business skills and the symbolic capital of entrepreneurship. This normalizes some practices, such as intensive networking within the oil and gas industry, while delegitimizing other practices like participating in public-sponsored entrepreneurial networking programs. This indicates that the importance of economic capital within Calgary's local field has displaced the symbolic capital of networking found within the field of technology entrepreneurship. Entrepreneurs both within and outside of the oil and gas sector placed little importance in status-building networking activities, preferring to spend that developing the business in other ways. Entrepreneurs' choice of networking practices emerge from this interaction between an entrepreneur's habitus and their understanding of the unwritten rules and norms of their local field.

Purposeful entrepreneurial networking in Waterloo, Ontario

The Waterloo Region, a municipality in Southwest Ontario made up of the cities of Waterloo, Kitchener, and Cambridge, is a center of Canadian innovation and entrepreneurship. The region's economy is propelled by the presence of both the headquarters of major international technology firms such as BlackBerry and OpenText as well as numerous smaller technology start-ups, many of which have received substantial venture investments and help constitute a successful entrepreneurial ecosystem (Spigel, 2015). There are two major reasons for the region's success as a hub of high-tech entrepreneurship despite its relatively low population. First, the University of Waterloo (UW), one of the world's leading institutions for computer science and engineering research, plays an important role in attracting skilled students to the region who then enter the workforce as entrepreneurs or skilled employees. The influence of UW dates back to its founding in 1957 as a polytechnic college designed to train engineers and technicians for the region's industrial economy (Bramwell, Nelles, & Wolfe, 2008; Nelles, Bramwell, & Wolfe, 2005). Second, the region has developed a local field that normalizes entrepreneurial activities and produces high rates of symbolic capital for high-tech entrepreneurship. Both these factors are aided by the presence of highly effective local economic development agencies which help reproduce the field and encourage particular practices, such as intensive networking between entrepreneurs. UW's development as an entrepreneurial university developed in tandem with this local culture and the larger field of technology entrepreneurship. This local culture has been reinforced by the successes of newer technology firms and the continued involvement of these firms' founders in promoting entrepreneurship, mentoring new founders, and as donors and supporters of the region's economic development and entrepreneurship organizations.

Beyond local factors, the region has been influenced in recent years by the norms and rules of the FTE. As argued above, this field normalizes activities such as risk-taking and the importance of social capital between entrepreneurs. This field influences the region through media channels like entrepreneurial and technology magazines and webpages, the presence of satellite offices of major international technology firms as well as by local economic development organizations encouraging these outlooks. One organization in particular, Communitech, has been very active in organizing programs that promote these values in both new and experienced technology entrepreneurs. Communitech hosts networking events, roundtable discussions for newer entrepreneurs to meet their more experienced counterparts, and runs incubator and accelerator facilities that provide subsidized office space and support for promising firms.

One of the most prominent outcomes of this overlapping field structure is the importance of networking between entrepreneurs in Waterloo. Interviewees in Waterloo drew upon an average of 2.61 out of 6 resources from their social network, compared to an average of 2.04 in Calgary. When lifestyle entrepreneurs – those with few plans for future growth or innovation – are excluded from the sample this difference increases to an average of 3.27 in Waterloo and 2.16 in Calgary. Similarly, 66 percent of respondents in Waterloo reported turning to their friends for business advice compared with 46 percent in Calgary. This suggests that networking, particularly with other entrepreneurs, is seen as an important part of the entrepreneurship process. The most popular view of networking in the region, as expressed by one entrepreneur is that: "here, unlike any other community that I've lived or worked [in], there's a strong sense of not just a desire, but a responsibility, to help up and coming companies, especially technology companies" (W114). This entrepreneur went on to observe that: "we do a good job of integrating people into the community and that builds strong ties ... I'd hazard a guess that we have more individuals in this community that have very broad, expansive networks than other communities." Local economic development officials had a similar view of regional networking, with one stating:

> what you'll find is that this community is very well networked. There are networking functions daily. I can go to any of those functions and know most of the people in that room. And we all circulate in different circles, but we're all connected.
>
> (W107)

Unlike in Calgary, entrepreneurs in Waterloo commonly viewed networking as an opportunity to develop business skills and learn how to solve business problems. In the words of one entrepreneur: "you find that you want to stay in touch with people just to see how you're doing compared to them. It's competitive, so you want to compare financials. Who's doing better this year?" (W110). Communitech's peer-to-peer events are an important platform for this kind of networking. Younger entrepreneurs often discussed these groups as a key tool in learning about the challenges of management, as one such entrepreneur described:

> when I say you can get anything that you want from Communitech, the truth is you can get anything you need from the community. Communitech

is the hub where you can access that. Peer to peers, you get to go and learn about problems other CEOs are having and how they're solving them.

(W117)

Networking with other entrepreneurs becomes an important part of identity creation amongst this group.

However, these views on networking were not universal within the community. Entrepreneurs who do not fit the region's vision of an ideal technology entrepreneur, due to their age, gender, or goals for the company, found it difficult to utilize the networking events organized by Communitech and other groups. For instance, the founder of a biotech firm noted that for him:

[Communitech is] not as valuable as it is for most companies in Kitchener-Waterloo. They have, at least until recently, been focused on the type of companies that usually grow up here, which are in [the] software area ... they don't really have much that can help us.

Similarly, another entrepreneur in the medial device market felt uncomfortable working with Communitech to find an advisor because he did not have a university education and instead turned to social contacts he made through his church for help. This exclusion is both direct, due to him not being welcomed in the networks since he will be seen as an "illegitimate" entrepreneur, as well as indirect as a result of his individual habitus. Many entrepreneurs who do not identify as growth-oriented technology entrepreneurs placed little value on the symbolic capital produced by participating Waterloo's entrepreneurial networks. Spending valuable time attending networking events or meeting with other entrepreneurs therefore makes little sense compared to spending that time building the business or with family and friends.

Discussion and conclusion

Calgary and Waterloo have very different local entrepreneurial fields with different relationships to the FTE. This has contributed to differences in networking practices found in the two communities. Calgary's local field has developed in tandem with the oil and gas industry rather than the FTE. This American-influenced industrial culture has contributed to the importance of entrepreneurship as a source of economic capital rather than symbolic capital. Calgary's local field positions entrepreneurship as a way to generate wealth rather than as a process of personal and technological development. As such, networking for its own sake has little value. The intensive networking found within the region's oil and gas cluster is aimed at finding new clients rather than helping participants build their entrepreneurial identity or skills. Calgary's field reduces the symbolic capital of this type of networking, lowering its importance relative to other entrepreneurial activities. This led to the normalization of particular types of networking practices, such as intensive networking with potential clients in the oil and gas industry while delegitimizing other practices such as attending networking events hosted by economic development organizations. Developing strong ties with other entrepreneurs is less important in Calgary because it does not contribute to the goals that are prioritized within the local field: creating a venture that can generate substantial windfall profits.

Waterloo's local field has been heavily influenced by the FTE through both the accidental cultural spillovers from UW as well as the active role of economic development agencies like Communitech. This led to the normalization of a particular form of entrepreneurship within the local field: the innovative, growth-oriented technology firm. The symbolic capital of producing a new venture based on a "cool" technology is highly prized in the field and is often judged as being more important than economic capital, at least in the short term. Entrepreneurs' networking practices tended to be organized around these norms. They reported spending more time networking and talking with other entrepreneurs in order to improve their business skills and learn from one another. They were much more willing to attend networking and peer groups organized through Communitech as well as engage in informal conversations with one another. The social capital this produced helps increase the symbolic capital of their entrepreneurial endeavors and increases their social status within the community.

Networking is a social activity. Though networking between entrepreneurs, advisors, clients, and customers is a crucial part of the entrepreneurship process, it is not a disembodied economic activity. The choice to network, with whom, and how much energy to dedicate to that activity as opposed to others, is deeply bound up in how individual actors understand the purpose of entrepreneurship and their relationships to a variety of contextual factors. These choices are made within the context of two major "fields": the local field and the larger field of technology entrepreneurship. The FTE spreads a global message of intensive networking as a natural part of the entrepreneurship process. Networking is a key way to establish a new venture's legitimacy and, more importantly, of developing an entrepreneurial identity. However, the FTE is interpreted through the social rules and outlooks of the local field that have developed out of a region's economic and social history. This contributes to the growth of distinctive patterns of entrepreneurial networking within different regions.

However, to simply attribute the development of these regional patterns on the local field does a disservice to the ability of entrepreneurs to strategically experiment with new practices to achieve their economic and social goals. From a Bourdieuian perspective, networking is a social practice developed within the context of the fields an entrepreneur operates in. Networking is not necessarily done purposefully to increase profits or help create legitimacy, but rather because particular networking practices make sense given the entrepreneur's habitus-based understanding of the field of technology entrepreneurship and their own local field. While seemingly an obvious point, this has several implications for the way researchers understand entrepreneurial networking. Understanding networking practices is often more important than understanding the size or density of entrepreneurs' networks. Entrepreneurs' goals when networking and their underlying belief about its important will inform both how they form networks and the types of knowledge and resources they can draw from the networks they build. Entrepreneurial networking is therefore a heterogeneous activity that develops within unique regional, industrial, and cultural contexts. Attempts to stimulate entrepreneurial networking in a region therefore have to take into account how networking is perceived within the local field and work within those constraints rather than trying to introduce a "Silicon Valley"-style perspective of entrepreneurial networking.

Note

1 The sectors were: computer and peripheral equipment manufacturing (NAICS 33411), software publishers (51121), data processing, hosting and related services (51821), computer systems design and related services (54141), scientific and technical consulting services (54169), and engineering services (54133).

References

Aldrich, H., & Zimmer, C. (1986). Entrepreneurship through social networks. In H. Aldrich (Ed.), *Population Perspectives on Organization* (pp. 13–28). Upsala: Acta Universistaits Upsaliensis.

Anderson, A. R., & Jack, S. L. (2002). The articulation of social capital in entrepreneurial networks: A glue or lubricant? *Entrepreneurship and Regional Development*, 14, 193–210.

Anderson, A. R., & Miller, C. J. (2003). "Class matters": human and social capital in the entrepreneurial process. *The Journal of Socio-Economics*, 32, 17–36.

Arenius, P., & de Clercq, D. (2005). A network-based approach on opportunity recognition. *Small Business Economics*, 24, 249–265.

Bourdieu, P. (1977). *Outline of a theory of practice*. Cambridge: Cambridge University Press.

Bourdieu, P. (1986). The forms of capital. In J. Richardson (Ed.), *Handbook of Theory and Research for the Sociology of Education* (pp. 241–258). New York: Greenwood.

Bourdieu, P. (1990). *The Logic of Practice*. Stanford: Stanford University Press.

Bourdieu, P. (2005). *The Social Structures of the Economy*. Cambridge: Polity.

Bourdieu, P., & Wacquant, L. J. D. (1992). *An Invitation to Reflexive Sociology*. Chicago: University of Chicago Press.

Bramwell, A., Nelles, J., & Wolfe, D. A. (2008). Knowledge, innovation and institutions: Global and local dimensions of the ICT cluster in Waterloo, Canada. *Regional Studies*, 42, 101–116.

Brüderl, J., & Preisendörfer, P. (1998). Network support and the success of newly founded businesses. *Small Business Economics*, 10, 213–225.

Burt, R. S. (2004). Structural holes and good ideas. *American Journal of Sociology*, 110, 349–399.

Centner, R. (2008). Places of privileged consumption practices: Spatial capital, the dot-com habitus, and San Francisco's Internet boom. *City & Community*, 7, 193–223.

Chastko, P. (2004). *Developing Alberta's Oil Sands: From Karl Clark to Kyoto.* Calgary: University of Calgary Press.

Conference Board of Canada (2012). *Metropolitan Forecast: GDP at Basic Prices by Industry.* Toronto.

de Clercq, D., & Voronov, M. (2009a). The role of cultural and symbolic capital in entrepreneurs' ability to meet expectations about conformity and innovation. *Journal of Small Business Management*, 47, 398–420.

de Clercq, D., & Voronov, M. (2009b). The role of domination in newcomers' legitimation as entrepreneurs. *Organization*, 16, 799–827.

de Clercq, D., & Voronov, M. (2009c). Towards a practice perspective of entrepreneurship: Entrepreneurial legitimacy as habitus. *International Small Business Journal*, 27, 395–419.

Dodd, S. D., & Anderson, A. R. (2007). Mumpsimus and the mything of the individualistic entrepreneur. *International Small Business Journal*, 25, 341–360.

Fligstein, N., & McAdam, D. (2012). *A Theory of Fields.* New York: Oxford University Press.

Foster, P. (1979). *The Blue-Eyed Sheiks: The Canadian Oil Establishment.* Toronto: Collins.

Gill, R., & Larson, G. S. (2014). Making the ideal (local) entrepreneur: Place and the regional development of high-tech entrepreneurial identity. *Human Relations*, 67, 519–542.

Granovetter, M. S. (1973). The strength of weak ties. *American Journal of Sociology*, 78, 1360–1381.

Henry, N., & Pinch, S. (2001). Neo-Marshallian nodes, institutional thickness, and Britain's "Motor Sport Valley": Thick or thin? *Environment and Planning A*, 33, 1169–1183.

Hoang, H., & Antoncic, B. (2003). Network-based research in entrepreneurship: A critical review. *Journal of Business Venturing*, 18, 165–187.

House, J. D. (1980). *The Last of the Free Enterprisers: The Oilmen of Calgary.* Toronto: Macmillan.

Jack, S. L. (2005). The role, use and activation of strong and weak network ties: A qualitative analysis. *Journal of Management Studies*, 42, 1233–1259.

James, A. (2005). Demystifying the role of culture in innovative regional economies. *Regional Studies*, 39, 1197–1216.

Karataş-Özkan, M. (2011). Understanding relational qualities of entrepreneurial learning: Towards a multi-layered approach. *Entrepreneurship and Regional Development*, 23, 877–906.

Klyver, K., & Foley, D. (2012). Networking and culture in entrepreneurship. *Entrepreneurship and Regional Development*, 24, 561–588.

Klyver, K., Hindle, K., & Meyer, D. (2008). Influence of social network structure on entrepreneurship participation: A study of 20 national cultures. *International Entrepreneurship and Management Journal*, 4, 331–347.

Lang, R., Fink, M., & Kibler, E. (2014). Understanding place-based entrepreneurship in rural Central Europe: A comparative institutional analysis. *International Small Business Journal*, 32, 204–227.

Lechner, C., & Dowling, M. (2003). Firm networks: External relationships as sources for the growth and competitiveness of entrepreneurial firms. *Entrepreneurship and Regional Development*, 15, 1–26.

Lucas, M., Sands, A., & Wolfe, D. A. (2009). Regional clusters in a global industry: ICT clusters in Canada. *European Planning Studies*, 17, 189–209.

Neff, G., Wissinger, E., & Zukin, S. (2005). Entrepreneurial labor among cultural producers: "Cool" jobs in "hot" industries. *Social Semiotics*, 15, 307–334.

Nelles, J., Bramwell, A., & Wolfe, D. A. (2005). History, culture and path dependency: Origins of the Waterloo ICT cluster. In D. A. Wolfe & M. Lucas (Eds.), *Global Networks and Local Linkages: The Paradox of Development in an Open Economy* (pp. 227–252). Montreal and Kingston: McGill-Queen's University Press.

Powell, W., White, D. R., Koput, K. W., & Owen-Smith, J. (2005). Network dynamics and field evolution: The growth of interorganizational collaboration in the life sciences. *American Journal of Sociology*, 110, 1132–1205.

Rutten, R., Westlund, H., & Boekema, F. (2010). The spatial dimension of social capital. *European Planning Studies*, 18, 863–871.

Saxenian, A. (1994). *Regional Advantage: Culture and Competition in Silicon Valley and Route 128*. Cambridge, MA: Harvard University Press.

Shane, S., & Cable, D. (2002). Network ties, reputation, and the financing of new ventures. *Management Science*, 48, 364–381.

Sorenson, O., & Stuart, T. (2001). Syndication networks and the spatial distribution of venture capital investments. *The American Journal of Sociology*, 106, 1546–1588.

Spigel, B. (2013). Bourdieuian approaches to the geography of entrepreneurial cultures. *Entrepreneurship & Regional Development*, 25, 804–818.

Spigel, B. (2015). The relational organization of entrepreneurial ecosystems. *Entrepreneurship Theory and Practice*.

Statistics Canada (2011). *Census of Canada*, 2011: Labour [computer file].

Stringfellow, L., Shaw, E., & Maclean, M. (2014). Apostasy versus legitimacy: Relational dynamics and routes to resource acquisition in entrepreneurial ventures. *International Small Business Journal*, 32, 571–592.

Stuart, T., & Sorenson, O. (2003). The geography of opportunity: Spatial heterogeneity in founding rates and the performance of biotechnology firms. *Research Policy*, 32, 229–253.

Swartz, D. (1997). *Culture and Power: The Sociology of Pierre Bourdieu*. Chicago: University of Chicago Press.

Thomson Reuters (2013). Special Tabulation: Private Equity and Venture Capital.

Uzzi, B. (1996). The sources and consequences of embeddedness for the economic performance of organizations: The network effect. *American Sociological Review*, 61, 674–698.

4 The persistence of regional entrepreneurship

Are all types of self-employment equally important?

Michael Fritsch and Michael Wyrwich

Introduction[1]

Empirical research on the persistence of regional entrepreneurship has begun only recently.[2] These analyses suggest that spatial differences in the level of entrepreneurship are rather robust and long-lasting. In a recent assessment of this phenomenon in Germany (Fritsch & Wyrwich, 2014), we have shown that there is persistence over the period from 1925 to 2005 despite several economic and political shocks such as the devastating World War II, occupation by the Allied forces and reconstruction of the country. Moreover, the Eastern part of Germany was under a socialist regime for 40 years that made intensive attempts to extinguish private sector economic initiative; after the socialist period, East Germany was subject to a shock transformation into a market economic system (see Fritsch, Bublitz, Sorgner, & Wyrwich, 2014, for a detailed analysis of the East German case). Since these patterns can hardly be explained by a persistence of the determinants of entrepreneurship, they may be regarded to indicate different regional traditions or "cultures" of entrepreneurship.

In this chapter we investigate the role of different kinds of historical self-employment in German regions on current levels of new business formation. We find that not all categories of self-employment in the year 1925 explain current regional levels of entrepreneurship equally. While, for example, we find a relatively strong effect for self-employment in non-agricultural private sector industries and in knowledge-intensive industries on current entrepreneurship, this effect is rather small or negligible for the self-employment rate that includes self-employed farmers, for homeworkers producing on their own account, as well as for self-employed women. We derive hypotheses that may explain the diverging importance of these types of self-employment for a persistent regional entrepreneurship culture.

Evidence on the persistence of regional entrepreneurship

A number of studies have found that the regional level of new business formation tends to be rather constant over time (e.g. Andersson & Koster

2011; Fotopoulos, 2014; Fritsch & Mueller, 2007; Fritsch & Wyrwich, 2014). One straightforward explanation for this pattern could be that regional determinants of new business formation are only changing slightly over time. Indeed, regional characteristics that are positively related to entrepreneurship, such as the qualification of the regional workforce or the employment share in small and young firms (e.g. Fritsch & Falck, 2007; Sutaria & Hicks 2004; Wagner, 2004), tend to be relatively stable over time periods of 10 to 20 years (Fotopoulos, 2014). Another explanation for the pronounced persistence of regional entrepreneurship may be found in different regional cultures (e.g. Andersson & Koster, 2011; Fritsch & Wyrwich, 2014). Entrepreneurial culture may be particularly relevant for explaining persistence of regional patterns in the case of Germany where a considerable number of economic and political shocks took place over the course of the twentieth century (Fritsch & Wyrwich, 2014).

A number of different definitions of what an entrepreneurial culture is can be found in the literature. It is, for example, described as a "positive collective programming of the mind" (Beugelsdijk, 2007, p. 190) or an "aggregate psychological trait" (Freytag & Thurik, 2007, p. 123) of the local population. A further characteristic of regions with a well-developed entrepreneurial culture could be that values such as individualism, independence, and achievement (e.g. Hofstede & McCrae, 2008; McClelland, 1961) are widespread among the inhabitants. A culture of entrepreneurship can be regarded as an informal institution that comprises norms, values, and codes of conduct (Baumol, 1990; North, 1994). Regions with a pronounced entrepreneurial culture should be characterized by a high level of social acceptance or "legitimacy" of entrepreneurship (Etzioni, 1987; Kibler, Kautonen, & Fink, 2014). They should also have a high share of persons with pronounced entrepreneurial personality traits such as extraversion, openness to experience and conscientiousness, as well as with a high ability to bear risk (Obschonka, Schmitt-Rodermund, Gosling, & Silbereisen, 2013; Rauch & Frese, 2007; Zhao & Seibert, 2006). A number of studies provide evidence that a culture of entrepreneurship can vary substantially across space within countries that have a uniform framework of formal institutions.[3]

Given the observation that informal institutions only change very slowly (North, 1994; Williamson, 2000), a culture of entrepreneurship should be rather long-lasting. In contrast to the persistence of informal institutions, formal institutions such as property rights, governance structures, and modes of resource allocation may change rather rapidly – although they are embedded in the informal institutions.[4] In a recent investigation of entrepreneurship in Germany, we found compelling evidence for the persistence of regional entrepreneurship over a period of 80 years from 1925 to 2005 (Fritsch & Wyrwich, 2014). Since this period was characterized by a number of radical changes of the political-economic environment, the persistence found cannot be caused by stability of the determinants of entrepreneurship but may be regarded to indicate the effect of an entrepreneurship culture.

The reasons for the persistence of a regional entrepreneurship culture are still unclear. One possible explanation could be the presence of entrepreneurial role models and the transmission of a positive entrepreneurial attitude in the regional population across generations (Laspita, Breugst, Heblich, & Patzelt, 2012). Entrepreneurial role models provide a non-pecuniary externality that reduces ambiguity and influences the decision to pursue an entrepreneurial career (Minniti, 2005). Furthermore, observing active entrepreneurs, especially successful ones, may increase social acceptance of entrepreneurship and self-confidence of people in regard to their ability to successfully set up their own business (Bosma, Hessels, Schutjens, van Praag, & Verheul, 2012; Kibler et al., 2014; Stuart & Sorenson, 2003). In this way, entrepreneurial role models may establish and reinforce a regional culture of entrepreneurship.

Figure 4.1 shows the interplay of these factors that may lead to a self- perpetuation of regional levels of entrepreneurship. It describes the virtuous circle by which high levels of new business formation lead to large numbers of entrepreneurial role models that may, in turn, trigger social acceptance of entrepreneurship over time. The high level of acceptance or "legitimacy" (Etzioni, 1987; Kibler et al., 2014) of entrepreneurship may stimulate entrepreneurial intentions among the population and the likelihood that people will decide to follow an entrepreneurial career path – again resulting in high start-up rates. Over time, the high levels of entrepreneurial activity and the social acceptance of entrepreneurship can also have an effect on the regional environment. A high demand for finance and advice may, for example, induce the emergence of a supporting infrastructure for entrepreneurship in fields such as finance and consulting services for newly founded businesses. Moreover, because most start-ups remain small, regions with high levels of new business formation tend to have many small firms that that are well known for functioning as a fruitful seedbed for start-ups (Parker, 2009).

Why should different forms of entrepreneurship matter for persistence?

While it appears plausible that historical rates of entrepreneurship indicate the extent of a regional culture of entrepreneurship, there are a number of arguments suggesting that not all types of entrepreneurship may represent this entrepreneurship culture to the same degree. One may, for example, claim that successful entrepreneurs whose ventures are opportunity based are more likely to induce positive role model and peer effects in the sense of "if they can do it, I can too" (Sorenson & Audia, 2000, p. 443) than entrepreneurs who are in business mainly out of necessity. Furthermore, self-employment in certain industries may be more significant for the perpetuation of an entrepreneurial culture based on role models and social acceptance of entrepreneurship than self-employment in other industries.

In order to throw more light on these issues we extend our earlier analysis (Fritsch & Wyrwich, 2014) by distinguishing between different forms

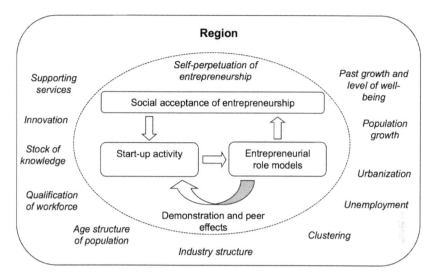

Figure 4.1 Determinants of new business formation and channels of self-perpetuation of regional entrepreneurship

of self-employment in the year 1925. One distinction is made between general self-employment and self-employment in non-agricultural private sector industries. The idea behind this categorization is that self-employment in non-agricultural parts of the economy is related to industrialization and economic development. It should therefore reflect more positively perceived role models and should be more closely associated with the generation of additional entrepreneurial opportunities than general self-employment that includes entrepreneurship in agriculture and in semi-private industries.[5] In a similar vein, entrepreneurship in knowledge-intensive industries may particularly foster the prevalence of positive entrepreneurial role models if entrepreneurial activities in these industries spur regional development. Moreover, Entrepreneurship in knowledge-intensive industries may generate relatively large numbers of further entrepreneurial opportunities. For these reasons we expect:

> *H1: Self-employment in non-agricultural private sector industries plays a more important role for the persistence of entrepreneurship than self-employment in general.*
> *H2: Self-employment in knowledge-intensive industries plays a more important role for the persistence of entrepreneurship than in non-knowledge-intensive industries.*

There are two types of self-employment in our historical data that can be assumed to include relatively large shares of necessity-driven entrepreneurship. These are self-employment by women and "domestic self-employment" by home workers (*Heimgewerbetreibende*). The role of women in the economic, political, and legal spheres of Imperial Germany (until 1918) and also in the Weimar Republic of the 1920s, was rather marginal. The Code Civil that became effective in the nineteenth century is especially revealing and stipulates that all important decisions within marriage are made by the husband. Accordingly, women had the "legal and economic status of social outsiders" (Schaser, 2008, p. 147). Women's legal position slowly improved in the 1920s, but nevertheless informal gender roles were largely persistent. Accordingly, self-employment by women did not reflect a socially well-accepted behavior around this time and is quite likely to have been mainly driven by necessity. Becker (1937), for instance, reports that many self-employed women in the 1920s were widows whose husbands had died throughout World War I. Given that self-employment by women reflects primarily necessity entrepreneurship, the prevalence of self-employed women in the early twentieth century should be only weakly or not at all related to the persistence of entrepreneurship.

Another form of primarily necessity-driven entrepreneurship in a historical perspective relates to self-employed homeworkers. This group performed narrowly defined tasks at home for just one or for very few firms and fall midway between an independent entrepreneur and a dependent employee (Statistik des Deutschen Reichs, 1927, p. 11). In the contemporaneous context, one can characterize this group as the dependent self-employed (Roman, Congregado, & Millàn, 2011). This pronounced dependence is a distinguishing feature of homeworkers and reflects a kind of necessity entrepreneurship. Accordingly, the prevalence of homeworkers in the early twentieth century should be only weakly or not at all related to the persistence of entrepreneurship. Based on these considerations, the following hypotheses regarding necessity-driven entrepreneurship can be formulated:

> *H3: The historical prevalence of self-employment by women plays a marginal role for persistence in entrepreneurship as compared to historical self-employment by men.*
> *H4: The historical level of homeworking plays a marginal role for persistence in entrepreneurship as compared to historical non-domestic self-employment.*

Data

Our empirical analysis is based on data on current start-up activity and historical self-employment rates in German regions. The data on new business formation are drawn from the Establishment History File of the Institute of Employment Research (IAB, Nuremberg). This dataset contains every establishment in Germany that employs at least one person obliged to make social insurance contributions (Gruhl, Schmucker, & Seth, 2012; Hethey & Schmieder, 2010). Establishments without any employees, i.e. solo-self-employment ventures, are not included in this data. The currently available information from the Establishment History Panel covers the years 1975 to 2010 and includes identifying start-ups from 1976 onwards. Information about the historical levels of entrepreneurship is taken from an extensive census of the year 1925 (Statistik des Deutschen Reichs, 1927). This historical data includes detailed information on the number of employees by gender, by 26 industries, as well as by "social status" at the level of counties (*kleinere Verwaltungsbezirke*). The variable "social status" distinguishes between blue-collar workers, white-collar employees, self-employed, homeworkers, and helping family members.

Our indicator for current levels of entrepreneurship is the average start-up rates for the time period of 1976 to 2010. We use three alternative definitions of this rate. The first of these definitions is the number of newly founded businesses over total employment in the region including employees in the public sector. A second definition is the number of newly founded businesses over the number of private sector employees. The reason for using these two definitions is that the number of private sector employees in the denominator may be influenced by a historically high level of entrepreneurship which is to a lesser degree the case for total employment. The third definition, the sector-adjusted start-up rate, accounts for differences of the regional industry structures and respective industry-specific entry conditions.[6]

Although the definition of administrative districts in the 1920s was different from what is defined as a district today, it is nevertheless possible to assign the historical districts to current planning regions, which represent functionally integrated spatial units comparable to labor market areas in the United States. If a historical district is located in two or more current planning regions, we assigned the employment to the respective planning regions based on each region's share of the geographical area. Thus, our regression framework is based on the 70 planning regions of West Germany.[7] Figure 4.2 shows the spatial distribution of the self-employment rate in the non-agricultural sector of the economy in the year 1925, Figure 4.3 depicts the regional share of homeworkers in this year, and the average regional start-up rate of the 1976–2010 period is shown in Figure 4.4.

In 1925, the self-employment rate in non-agricultural industries was particularly high in the northern part of Germany, in the south around Munich, and in certain regions of Baden-Wuerttemberg in the southwest, whereas the Ruhr area (north of Cologne) was marked by relatively low rates of self-employment (Figure 4.2). The homeworker rate in 1925 shows a quite different spatial pattern (Figure 4.3) with rather high rates around Stuttgart and in regions close to Nuremberg. The average start-up activity today (Figure 4.4) shows some considerable correspondence with the historical

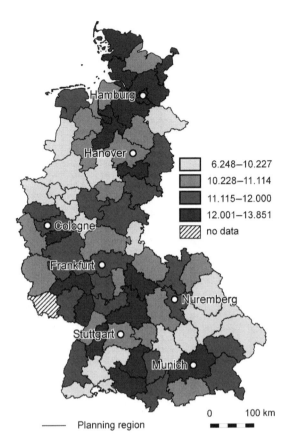

Figure 4.2 Self-employment rate in non-agricultural industries 1925
Source: Statistik des Deutschen Reichs, 1927.

level of self-employment in non-agricultural industries in 1925, but there are also some notable deviations such as the low start-up rates in most parts of Baden-Wuerttemberg.

In order to test the effect of the different types of historical entrepreneurship on the level of new business formation in the years 1976 to 2010, we regress the average yearly start-up rates for this period on different definitions of the historical self-employment rates including a number of control variables. The overall level of self-employment is given by the overall number of self-employed divided by the total number of employees. The self-employment in the non-agricultural private sector is the number of self-employed in manufacturing and services over all employees. We also test the effect of the self-employment rate of men and woman and the rate of homeworkers. Self-employed in "machine, apparatus, and vehicle

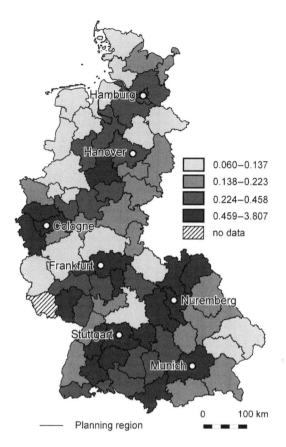

Figure 4.3 Home worker rate in non-agricultural industries 1925
Source: Statistik des Deutschen Reichs, 1927.

construction," "electrical engineering, precision mechanics, optics," "chemical," and "rubber- and asbestos" are regarded as knowledge-intensive. The numbers of self-employed in these industries are divided by the total number of employees.

As control variables for regionally different conditions for entrepreneurship we use the share of manufacturing employment in 1925, the population share of expellees in 1950, as well as population density and the share of R&D employees[8] in the year 1976. The population share of expellees is taken from the 1950 Census and is included to account for the massive inflow of people from outside the region after World War II that might have had an influence on the regional culture of entrepreneurship. Table 4.1 provides an overview on the definition of the variables used in the analysis.[9]

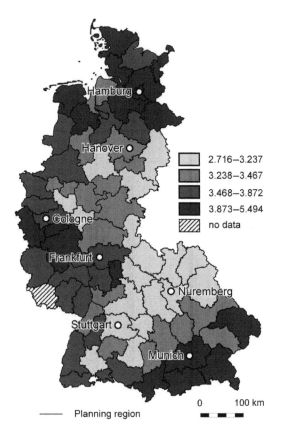

Figure 4.4 Average start-up rate 1976–2010
Source: Social Insurance Statistics.

Empirical analysis

We begin our analysis with the 1925 overall self-employment rate (Table 4.2) and then move on to other definitions of self-employment (Tables 4.3 to 4.5). Regressing the entire share of people that have been counted as self-employed in 1925 on the average start-up rate of the period from 1976 to 2010 (Table 4.2) shows no robust significant relationship. Restricting the definition of entrepreneurship to the non-agricultural private sector, we find a positive effect in all versions of the model, i.e. regardless of how far regional differences in industry structures and entry conditions are accounted for (Table 4.3). These results are in accordance with our Hypothesis H1.[10]

In the models presented in Tables 4.2 and 4.3 the specialization in manufacturing industries in 1925 that controls for sector-specific effects at that time is negatively related to the level of start-up activity in the period from 1976

Table 4.1 Definition of variables

Variable	Definition
Average start-up rate 1976–2010 (total employment)	Number of start-ups in a region per year over the total number of employees.[b]
Average start-up rate 1976–2010 (private sector employment)	Number of start-ups in a region per year over total private sector employment.[b]
Average sector adjusted start-up rate 1976–2010	Sector adjusted number of start-ups in a region over total employment.[b]
Self-employment rate 1925	Total number of self-employed persons divided by the total number of employees.[a]
Self-employment rate in non-agricultural industries 1925	Number of self-employed persons in non-agricultural private industries divided by the total number of employees.[a]
Self-employment rate females in non-agricultural industries 1925	Number of self-employed women in non-agricultural private industries divided by all working women.[a]
Self-employment rate males in non-agricultural industries 1925	Number of self-employed men in non-agricultural private industries divided by all working men.[a]
Homeworker rate in non-agricultural industries 1925	Number of persons registered as homeworkers in non-agricultural private industries divided by the total number of employees.[a]
Share of self-employed women over total self-employment 1925	Number of self-employed women in in non-agricultural private industries over the total number of persons registered as self-employed in these industries.[a]
Self-employment rate in non-agricultural industries (incl. homeworkers) 1925	Number of self-employed persons and number of homeworkers in non-agricultural private industries divided by the total number of employees.[a]
Share of homeworkers over total self-employment 1925	Number of homeworkers 1925 in non-agricultural private industries over all self-employed and homeworkers in non-agricultural private industries.[a]
Self-employment rate in knowledge intensive industries 1925	Number of self-employed persons in knowledge-intensive industries ("machine, apparatus, and vehicle construction," "electrical engineering, precision mechanics, optics," "chemical," and "rubber- and asbestos") divided by the total number of employees.[a]
Self-employment rate in non-knowledge-intensive industries 1925	Number of self-employed persons in non-knowledge-intensive industries divided by the total number of employees.[a]
Employment share in knowledge-intensive industries 1925	Number of employees in knowledge-intensive industries divided by the total number of employees.[a]

Table 4.1 (*cont.*)

Variable	Definition
Employment share in non-knowledge-intensive industries 1925	Number of employees in non-knowledge-intensive industries divided by the total number of employees.[a]
Employment share in manufacturing 1925	Number of employees in manufacturing divided by the total number of employees.[a]
Population share of expellees 1950	Number of expellees over regional population.[d]
Population density 1974	Number of inhabitants per km^2.[c]
Share of R&D employees 1976	Number of employees working as engineers and natural scientists divided by the total number of employees.[c]

Sources: [a] Statistik des Deutschen Reichs (1927); [b] Social Insurance Statistics; [c] Federal Statistical Office; [d] Census 1950 (various volumes). All variables enter the models in log-form.

Table 4.2 General self-employment rate 1925 and regional levels of start-up activity in the 1976–2010 period

	Number of start-ups divided by ...			Sector-adjusted start-up rate
	Total employment		Private sector employment	
Self-employment rate 1925	0.00569	−0.0120	−0.0879	0.00532
	(0.0360)	(0.146)	(0.167)	(0.0956)
Employment share in manufacturing 1925	-	−0.158**	−0.247***	−0.0222
		(0.0633)	(0.0650)	(0.0416)
Population share of expellees 1950	-	−0.108**	−0.138**	−0.0801**
		(0.0455)	(0.0535)	(0.0326)
Population density 1974	-	0.0390	0.0253	−0.0796*
		(0.0693)	(0.0825)	(0.0451)
Share of R&D employees 1976	-	−0.00189	0.00337	−0.0890***
		(0.0492)	(0.0512)	(0.0327)
Federal state dummies	No	Yes***	Yes***	Yes***
Constant	1.251***	0.542	0.654	1.075***
	(0.0581)	(0.459)	(0.504)	(0.283)
R^2	0.000	0.333	0.486	0.577

Notes: N=70. Robust standard errors in parentheses; ***: statistically significant at the 1 percent level; **: statistically significant at the 5 percent level; *: statistically significant at the 10 percent level. The independent variables (except the federal state dummies) are entered as log values.

to 2010. This effect is, however, only found to be statistically significant in models II and III. An explanation could be that regions with a high share of manufacturing employment in the year 1925 still have relatively high levels of manufacturing activities that are characterized by higher entry barriers than other sectors. There is indeed a high correlation between the share of employment in manufacturing in 1925 and the respective share today (r = 0.43 for the mid-1970s and r = 0.23 on average for the period from 1976 to 2010). That the effect becomes insignificant when the current sector adjusted start-up rate is taken as the dependent variable is not surprising since this version of the start-up rate already accounts for sector-specific effects. The population share of expellees shortly after World War II is negatively related to the start-up rate regardless of the sector-adjustment. The reason might be that expellees either had a relatively low propensity of starting an own business or have typically settled in regions with low entrepreneurial opportunities.[11] Population density and the share of R&D employees are only weakly or not at all related to the uncorrected start-up rate (model II), whereas there is a negative relationship in the model for the sector-adjusted rate (model IV).[12]

Table 4.3 Self-employment rate in the non-agricultural private sector 1925 and persistence of regional entrepreneurship

	Number of start-ups divided by …			Sector-adjusted start-up rate
	Total employment		Private sector employment	
Self-employment rate in non-agricultural industries 1925	0.244**	0.314***	0.294**	0.197***
	(0.107)	(0.103)	(0.134)	(0.0658)
Employment share in manufacturing 1925	-	−0.191***	−0.273***	−0.0441
		(0.0523)	(0.0633)	(0.0365)
Population share of expellees 1950	-	−0.118**	−0.142**	−0.0871**
		(0.0475)	(0.0547)	(0.0344)
Population density 1974	-	0.0789*	0.0934*	−0.0597**
		(0.0416)	(0.0503)	(0.0232)
Share of R&D employees 1976	-	−0.0376	−0.0306	−0.111***
		(0.0494)	(0.0533)	(0.0325)
Federal state dummies	No	Yes***	Yes***	Yes***
Constant	1.782***	0.760*	0.812*	1.220***
	(0.240)	(0.384)	(0.445)	(0.261)
R^2	0.055	0.386	0.510	0.603

Notes: N=70. Robust standard errors in parentheses: ***: statistically significant at the 1 percent level; **: statistically significant at the 5 percent level; *: statistically significant at the 10 percent level. The independent variables (except the federal state dummies) are entered as log values.

Distinguishing different types of self-employment in the non-agricultural private sector shows a significantly positive effect only for self-employed males (Table 4.4). Self-employment of females and homeworkers that probably comprises a high share of necessity entrepreneurship is not statistically significant. These results are in accordance with our Hypotheses H3 and H4.[13]

Finally, we investigate the role of entrepreneurship in knowledge-intensive industries (Table 4.5). We run a "horse race" in our regression models where we include self-employment in knowledge-intensive non-agricultural industries in 1925 along with the self-employment rate in non-knowledge-intensive non-agricultural private sector industries. Additionally, we include the employment shares of both sectors in 1925 in order to control for their relevance in the local economy. The results suggest especially that historical self-employment in knowledge-intensive industries can be regarded as

Table 4.4 Different forms of self-employment and persistence of regional entrepreneurship

	Number of start-ups divided by …			Sector-adjusted start-up rate
	Total employment		Private sector employment	
Self-employment rate males in non-agricultural industries 1925	0.218**	0.277***	0.275**	0.171***
	(0.0860)	(0.0803)	(0.117)	(0.0551)
Self-employment rate of females in non-agricultural industries 1925	0.0983	–0.0549	–0.0589	–0.00271
	(0.0838)	(0.128)	(0.152)	(0.0955)
Homeworker rate in non-agricultural industries 1925	–0.0340*	0.0112	0.00379	0.00309
	(0.0174)	(0.0222)	(0.0251)	(0.0148)
Employment share in manufacturing 1925	-	–0.245***	–0.311***	–0.0623
		(0.0733)	(0.0893)	(0.0452)
Population share of expellees 1950	-	–0.111**	–0.135**	–0.0851**
		(0.0482)	(0.0568)	(0.0337)
Population density 1974	-	0.0855**	0.101**	–0.0572**
		(0.0378)	(0.0485)	(0.0223)
Share of R&D employees 1976	-	–0.0120	–0.0109	–0.104***
		(0.0501)	(0.0597)	(0.0340)
Federal state dummies	No	Yes***	Yes***	Yes***
Constant	1.753***	0.653	0.678	1.177***
	(0.266)	(0.402)	(0.476)	(0.312)
R^2	0.117	0.406	0.522	0.608

Notes: N=70. Robust standard errors in parentheses: ***: statistically significant at the 1 percent level; **: statistically significant at the 5 percent level; *: statistically significant at the 10 percent level. The independent variables (except the federal state dummies) are entered as log values.

a driver of a persistent entrepreneurship culture, whereas self-employment in other private sectors is not significantly related to current start-up activity. These findings are in line with our Hypothesis H2. The results for the control variables in Tables 4.4 and 4.5 are in line with the former models presented in Tables 4.2 and 4.3. It should be noted that the historical employment share in knowledge-intensive industries is negatively related to current

Table 4.5 Knowledge intensity of industries and persistence of regional entrepreneurship

	Number of start-ups divided by …		Sector-adjusted start-up rate
	Total employment	Private sector employment	
Self-employment rate in knowledge-intensive industries 1925	0.280***	0.265**	0.189***
	(0.0624)	(0.110)	(0.0526)
Self-employment rate in non-knowledge-intensive industries 1925	–	-0.0648	0.00257
		(0.210)	(0.0929)
Employment share in knowledge-intensive industries 1925	-0.139***	-0.154***	-0.0801***
	(0.0340)	(0.0386)	(0.0220)
Employment share in non-knowledge-intensive industries 1925	–	0.204	0.175
		(0.199)	(0.109)
Population share of expellees 1950	-0.0921**	-0.135**	-0.0769**
	(0.0451)	(0.0516)	(0.0326)
Population density 1974	0.0586	-0.00406	-0.0989***
	(0.0376)	(0.0896)	(0.0370)
Share of R&D employees 1976	0.00644	0.0244	-0.0853**
	(0.0472)	(0.0558)	(0.0340)
Federal state dummies	No	Yes***	Yes***
Constant	1.942***	2.415**	2.192***
	(0.516)	(0.978)	(0.449)
R^2	0.443	0.506	0.686

Notes: N=70. Robust standard errors in parentheses: ***: statistically significant at the 1 percent level; **: statistically significant at the 5 percent level; *: statistically significant at the 10 percent level. The independent variables (except the federal state dummies) are entered as log values.

start-up activity. Since the respective industries are belonging to the manufacturing sector they are characterized by relatively high entry barriers. Due to industrial path-dependency, regions historically specialized in these industries might have specialized in industries with similar entry conditions which might explain the negative relationship. There is indeed a high correlation between the employment share in knowledge-intensive industries in 1925 and the share of R&D employment in 1976 ($r = 0.67$).

Conclusions

Recent research has shown that regional differences in entrepreneurial activities are characterized by pronounced long-term persistence. The empirical evidence suggests that a considerable part of this persistence can be attributed to the regional culture of entrepreneurship. Assuming that demonstration and role-model effects are a key mechanism for the self-perpetuating nature of such a culture, we have investigated the role of different types of self-employment for the persistence of regional entrepreneurship.

We find indeed that not all types of self-employment are equally important for the long-term persistence of regional entrepreneurship. While the overall level of self-employment that includes agriculture and semi-public sectors is only weakly related to the current level of new business formation, we find a rather pronounced positive effect for the share of self-employed males in non-agricultural industries as well as for self-employment in knowledge-intensive industries. Self-employment by women and the share of homeworkers that may both include large shares of necessity-driven entrepreneurship do not contribute to an explanation of persisting regional differences in the level of new business formation.

Our findings have a number of implications. First of all, not only the sheer number of start-ups and self-employed persons is important for the persistence of a regional entrepreneurship culture, but particularly those forms of entrepreneurship that can be assumed to be most relevant for growth. Hence, policies that aim at fostering a persistent culture of entrepreneurship may focus on those types of new businesses. Second, attempts to measure a regional culture should account for such differences and should therefore also focus on those types of entrepreneurship that have a longer-lasting effect. Third, the persistence of start-up activity shows that policy initiatives that aim at creating an entrepreneurial culture need a long-term orientation. Effects of such policies may become visible only after longer periods of time. However, once these effects become manifest, they may be long-lasting.

One important question that we have not mentioned yet is that the characteristics of historically-grown entrepreneurship cultures might differ across regions, and there may also be differences in the importance of the mechanisms by which such a culture persists over longer periods of time. To identify and analyze such differences could considerably enhance our understanding of the phenomenon. In this chapter we have attempted to make a step in this direction.

Notes

1 We are indebted to the editors for helpful comments on an earlier draft.
2 E.g. Andersson and Koster (2011), Fotopoulos (2014), Fritsch and Mueller (2007), Fritsch and Wyrwich (2014).
3 For example, Andersson (2012), Aoyama (2009), Beugelsdijk (2007), Davidsson (1995), Davidsson and Wiklund (1997), Etzioni (1987), Kibler, Kautonen, and Fink (2014), Rentfrow, Gosling, and Potter (2008), Westlund and Adam (2010), Westlund and Bolton (2003), and Westlund, Larsson, and Olsson (2014).
4 East Germany is a good example of the differences between formal and informal institutions. With the reunification of Germany in 1990, the ready-made West German framework of formal institutions became effective practically overnight. However, more than two decades later a specific East German mentality can still be identified.
5 In the semi-private industries it is not possible to clearly assign economic units to the private sector and the state based on the available historical dataset. One example is the "Transport and Communications" sector, which comprises several private but also many public companies and employment. "Self-employed" in these industries can be true entrepreneurs but also managers of state-owned units.
6 The sector-adjusted number of start-ups is defined as the number of new businesses in a region that would be expected if the composition of industries was identical across all regions. Thus, the measure adjusts the original data by making industry composition uniform across regions. The adjustment procedure is based on the regional distribution of 59 industries and follows a shift-share technique (Ashcroft, Love, & Malloy, 1991; the Appendix of Audretsch & Fritsch, 2002). We calculate the sector adjusted rates for every year of the 1976–2010 period and use the average values of these yearly rates.
7 There are 74 West German planning regions. For administrative reasons, the cities of Hamburg and Bremen are defined as planning regions even though they are not functional economic units. To avoid distortions, we merged these cities with adjacent planning regions. Hamburg is merged with the region of Schleswig-Holstein South and Hamburg-Umland-South. Bremen is merged with Bremen-Umland. Further, we exclude the planning region "Saarland" from the regression analysis since most of the areas within this planning region was not completely under German administration at the time of the 1925 census.
8 R&D employees are defined as those with tertiary degrees working as engineers or natural scientists. Population density is a "catch-all" variable for regional conditions that captures many of the region-specific conditions shown in Figure 4.1.
9 For descriptive statistics and correlations among variables see the Appendix of the Working Paper version of this article (http://zs.thulb.uni-jena.de/receive/jportal_jparticle_00342984).
10 Running the models with the self-employment rate in agriculture (self-employed farmers divided by the total number of employees) reveals that there is no significant relationship with current start-up activity.
11 The average regional self-employment rate among expellees in 1950 was 4.1 percent whereas the self-employment rate of the original local population was about 14.2 percent. Thus, given that the entrepreneurial propensity of expellees was indeed considerably lower, a high population share of expellees accordingly feeds back into relatively low start-up activity. Furthermore, there is a positive correlation of r = 0.43 between the population share of expellees and the self-employment

rate in 1950. This suggests that expellees were not particularly relocated in regions with below average levels of entrepreneurial activity

12 The insignificance of the share of R&D employees in the models for the start-up rate and the negative effect in the models for the sector-adjusted rate of new firm formation is somewhat surprising since we have quite stable and significant positive effects for this variable in earlier work (Fritsch & Wyrwich, 2014). This positive effect is confirmed when restricting our analysis to the more recent sub-period 1992 to 2010 and applying pooled regression. A possible reason for the differences in the results may be an increase of R&D activities in smaller firms over the last decades that might be more likely to induce start-up activity as compared to R&D activities in large firms. Unfortunately, we cannot test this conjecture with the dataset at hand.

13 We also run robustness check with the self-employment rate in non-agricultural industries 1925 and the share of homeworkers and self-employed women respectively that confirm the results of Table 4.4 (see Table A4 and A5 in the Appendix of the Working Paper version of this article (http://zs.thulb.uni-jena.de/receive/jportal_jparticle_00342984)).

References

Andersson, M. (2012). Start-up rates, entrepreneurship culture and the business cycle: Swedish patterns from national and regional data. In Pontus Braunerhjelm (Ed.), *Entrepreneurship, norms and the business cycle*, Swedish Economic Forum Report 2012, Stockholm: Entreprenörskapsforum, 91–110.

Andersson, M., & Koster, S. (2011). Sources of persistence in regional start-up rates: Evidence from Sweden. *Journal of Economic Geography*, 11, 179–201.

Aoyama, Y. (2009). Entrepreneurship and regional culture: The case of Hamamatsu and Kyoto, Japan. *Regional Studies*, 43, 495–512.

Ashcroft, B., Love, J. H., & Malloy, E. (1991). New firm formation in the British counties with special reference to Scotland. *Regional Studies*, 25, 395–409.

Audretsch, D. B., & Fritsch, M. (2002). Growth regimes over time and space. *Regional Studies*, 36, 113–124.

Baumol, W. J. (1990). Entrepreneurship: Productive, unproductive, and destructive. *Journal of Political Economy*, 98, 893–921.

Becker, U. (1937). *Die Entwicklung des Frauenerwerbs seit der Jahrhundertwende*. Bleicherode am Harz: Carl Piest.

Beugelsdijk, S. (2007). Entrepreneurial culture, regional innovativeness and economic growth. *Journal of Evolutionary Economics*, 17, 187–210.

Bosma, N., Hessels, J., Schutjens, V., van Praag, M., & Verheul, I. (2012). Entrepreneurship and role models. *Journal of Economic Psychology*, 33, 410–424.

Census (1950). *Ergebnisse der Volks- und Berufszählung vom 13. September 1950 in den Ländern der Bundesrepublik Deutschland*. Various volumes, Statistical Offices of the Federal States of Germany.

Davidsson, P. (1995). Culture, structure and regional levels of entrepreneurship. *Entrepreneurship and Regional Development*, 7, 41–62.

Davidsson, P., & Wiklund, J. (1997). Values, beliefs and regional variations in new firm formation rates. *Journal of Economic Psychology*, 18, 179–199.

Etzioni, A. (1987). Entrepreneurship, adaptation and legitimation. *Journal of Economic Behavior and Organization*, 8, 175–199.

Fotopoulos, G. (2014). On the spatial stickiness of UK new firm formation rates. *Journal of Economic Geography*, 14, 651–679.

Freytag, A., & Thurik, R. (2007). Entrepreneurship and its determinants in a cross-country setting. *Journal of Evolutionary Economics*, 17, 117–131.

Fritsch, M., & Falck, O. (2007). New business formation by industry over space and time: A multi-dimensional analysis. *Regional Studies*, 41, 157–172.

Fritsch, M., & Mueller, P. (2007). The persistence of regional new business formation-activity over time: Assessing the potential of policy promotion programs. *Journal of Evolutionary Economics*, 17, 299–315.

Fritsch, M., & Wyrwich, M. (2014). The long persistence of regional levels of entrepreneurship: Germany 1925 to 2005. *Regional Studies*, 48, 939–954.

Fritsch, M., Bublitz, E., Sorgner, A., & Wyrwich, M. (2014). How much of a socialist legacy? The re-emergence of entrepreneurship in the East German transformation to a market economy. *Small Business Economics*, 43, 427–446.

Gruhl, A., Schmucker, A., & Seth, S. (2012). The Establishment History Panel 1975–2010: Handbook version 2.2.1, FDZ-Datenreport, 04/2012, Nuremberg.

Hethey, T., & Schmieder, J. F. (2010). Using worker flows in the analysis of establishment turnover: Evidence from German administrative data. FDZ-Methodenreport 06-2010 EN, Research Data Centre of the Federal Employment Agency (BA) at the Institute for Employment Research (IAB): Nuremberg.

Hofstede, G., & McCrae, R. R. (2008). Personality and culture revisited, linking traits and dimensions of culture. *Cross-Cultural Research*, 38, 52–87.

Kibler, E., Kautonen, T., & Fink, M. (2014). Regional social legitimacy of entrepreneurship: Implications for entrepreneurial intention and start-up behaviour. *Regional Studies*, 48, 995–1015.

Laspita, S., Breugst, N., Heblich, S., & Patzelt, H. (2012). Intergenerational transmission of entrepreneurial intentions. *Journal of Business Venturing*, 27, 414–435.

McClelland, D. C. (1961). *The achieving society*. Princeton: Van Nostrand Reinhold.

Minniti, M. (2005). Entrepreneurship and network externalities. *Journal of Economic Behavior and Organization*, 57, 1–27.

North, D. C. (1994). Economic performance through time. *American Economic Review*, 84, 359–368.

Obschonka, M., Schmitt-Rodermund, E., Gosling, S. D., & Silbereisen, R.K. (2013). The regional distribution and correlates of an entrepreneurship-prone personality profile in the United States, Germany, and the United Kingdom: A socioecological perspective. *Journal of Personality and Social Psychology*, 105, 104–122.

Parker, S. (2009). Why do small firms produce the entrepreneurs? *Journal of Socio-Economics*, 38, 484–494.

Rauch, A., & Frese, M. (2007). Let's put the person back into entrepreneurship research: A meta-analysis on the relationship between business owners' personality traits, business creation, and success. *European Journal of Work and Organizational Psychology*, 16, 353–385.

Rentfrow, J. P., Gosling, S. D., & Potter, J. (2008). A theory of the emergence, persistence, and expression of geographic variation in psychological characteristics. *Perspectives on Psychological Science*, 3, 339–369.

Román, C., Congregado, E., & Millán, J. M. (2011). Dependent self-employment as a way to evade employment protection legislation. *Small Business Economics*, 37, 363–392.

Schaser, A. (2008). Gendered Germany. In J. Retallack (Ed.), *Imperial Germany: 1871–1918*, Oxford: Oxford University Press, 128–150.

Sorenson, O., & Audia, P. G. (2000). The social structure of entrepreneurial activity, geographic concentration of footwear production in the United States, 1940–1989. *American Journal of Sociology*, 106, 424–462.

Statistik des Deutschen Reichs (1927). *Volks-, Berufs- und Betriebszählung vom 16. Juni 1925: Die berufliche und soziale Gliederung der Bevölkerung in den Ländern und Landesteilen*, Vol. 403–Vol. 405, Berlin: Reimar Hobbing.

Stuart, T. E., & Sorenson, O. (2003). The geography of opportunity: Spatial heterogeneity in founding rates and the performance of bio-technology firms. *Research Policy*, 32, 229–253.

Sutaria, V., & Hicks, D. A. (2004). New firm formation: Dynamics and determinants. *Annals of Regional Science*, 38, 241–262.

Wagner, J. (2004). Are young and small firms hothouses for nascent entrepreneurship? Evidence from German micro data. *Applied Economics Quarterly*, 50, 379–391.

Westlund, H., & Bolton, R. E. (2003). Local social capital and entrepreneurship. *Small Business Economics*, 21, 77–113.

Westlund, H., & Adam, F. (2010). Social capital and economic performance: A meta-analysis of 65 studies. *European Planning Studies*, 18, 893–919.

Westlund, H., Larsson, J. P., & Olsson, A. R. (2014). Startups and local social capital in Swedish municipalities. *Regional Studies*, 48, 974–994.

Williamson, O. (2000). The new institutional economics: Taking stock, looking ahead. *Journal of Economic Literature*, 38, 595–613.

Zhao, H., & Seibert, S. E. (2006). The big-five personality dimensions and entrepreneurial status: A meta-analytical review. *Journal of Applied Psychology*, 91, 259–271.

5 Two strands of entrepreneurship

A tale of technology ventures and traditional small businesses in South Korea

Jun Koo and Songhee Yoo

Introduction

The entrepreneurship literature has shown that new firm formation is an important source of job creation and regional development (Birch, 1981; Davisson, Lindmark, & Olofsson, 1994; Fritsch & Mueller, 2004; Mueller, Van Stel, & Storey, 2008). In particular, recent studies pay particular attention to the role of technology-oriented new ventures, such as Google and Facebook (Acs, Audretsch, & Carlsson, 2003; Shane, 2001). However, there is another kind of start-up, often overlooked because they are not as glamorous as technology firms. People often start a new business to make ends meet. Small restaurants, mom-and-pop retail stores, and small business services are cases in point. In addition, small traditional manufacturing firms can also be grouped in this category. They are often home-based with a purpose of income substitution. Contrary to popular belief, this kind of new business comprises a lion's share of new firm formations in Korea (KISED, 2013). Korea is often considered an innovation-driven entrepreneurial economy because of a strong image created by high-tech firms, such as Samsung and LG. However, technology-oriented entrepreneurial activities are quite limited in Korea, as we will demonstrate below. On the other hand, Korea has over six times as many restaurants and four times as many retail stores per capita as the United States has. In addition, their growth rates have been much greater than that of any other sector in Korea. Therefore, the existing focus on technology ventures may grossly distort the picture of new firm activities and their roles in regional development in Korea or possibly in other countries at a similar development stage.

Against this background, this study examines the geographic distribution of both technology ventures and traditional small businesses in Korea. A comparison between the two types of new firms will reveal different growth potentials of the various regions. In particular, start-up statistics show that technology ventures have little presence in most regions in Korea and that entrepreneurial activities are predominantly traditional small businesses. This study also aims to evaluate entrepreneurial policies at both the central and local levels in Korea. Technology ventures and traditional small businesses may look alike in start-up statistics. However, they are markedly different in nature and need tailored policy approaches. Accordingly, policy makers should pay more attention to developing a customized policy framework for the different types of new firms. However, our analysis reveals that entrepreneurial policy at both local and central levels in Korea has a strong focus on technology ventures, thereby creating a significant policy mismatch.

For the analysis, this study relies upon start-up statistics and surveys compiled by the Korea Institute of Startup and Entrepreneurship Development (KISED). In addition to the existing secondary data, we also conducted in-depth telephone interviews with ten entrepreneurs who recently started their businesses. The interviewees were selected from a list of entrepreneurs compiled by the Startup Network, a government-sponsored information portal which is designed to provide information regarding starting a new business. In selecting the interviewees, we considered sectoral and geographical distributions, resulting in five technology ventures and five traditional small businesses located both in the capital and non-capital regions. The interviews primarily focused on motivation, obstacles, and policy support for entrepreneurship and are used as anecdotal evidence to support our argument.[1]

Role of new firms in regional development

New firms have been considered a driving force of regional development for several reasons. First, they are an important source of job creation. Since Birch's seminal works (Birch, 1981, 1987), a plethora of research has reported that new firms account for a significant share of job creation (Broersma & Gautier, 1997; Fritsch & Mueller, 2004; Picot & Dupuy, 1998; Van Praag & Versloot, 2007). For instance, Verhoeven et al. (2005) reported that new firms accounted for some 30 percent of the total job creation in the Netherlands. De Kok et al. (2006) also showed that two-thirds of net annual employment growth can be attributable to small firms. In particular, the job contribution of new firms is reported to be stronger in the case of technology ventures compared with traditional small businesses (Baptista & Preto, 2011).

Second, new firms are often considered hotbeds for innovation (Van Praag & Versloot, 2007). New firms play an important role not only in producing new technology but also in spilling it over to other firms in proximity. Moreover, an incumbent firm can be stimulated to imitate or innovate, facing stiff competition with local new firms (Van Stel & Suddle, 2008). Such effects are often believed to create geographically bounded technology spillovers (Acs et al., 2003; Feldman, 1994). This line of argument has gained increasing popularity with a growing emphasis on technology and innovation in the so-called knowledge economy. In addition, it is also often argued that the new firms' contribution to innovation serves as an important driving force for regional productivity growth (Aghion, Blundell, Griffith, Howitt, & Prantl, 2004; Disney, Haskel, & Heden, 2003).

Although a lion's share of the existing literature supports the positive role of new firms (technology-oriented ones in particular) in regional development, some scholars remain skeptical about new firms as an engine of job creation and innovation. Davis et al. (1996) showed that the net job creation effect attributable to new firms is overestimated because of the high job destruction rate among small firms. Haltiwanger et al. (1999) found that typical new firms in the United States are less productive than existing firms, implying that their growth potential is not that rosy. More recently, Shane (2009) argued that the majority of new start-ups in the United States are not real entrepreneurial firms in that they do not aim to grow. Drawing upon data from the U.S. Bureau of Labor Statistics, he also found that new firms accounted for only 7 percent of the new jobs created in 2004. Shane made a clear point that the new firms' contribution to job creation should be considered a myth.

This discrepancy existing in the literature can be attributable to the focus of analysis. The majority approach in the literature focuses on a subset of new firms with high growth potential. Most new firms in this category fall in high-tech sectors, such as IT, biotech, or knowledge-oriented business services. These types of new firms account for only a fraction of all new firms but are responsible for a disproportionately large amount of job and wealth creation (Henrekson & Johansson, 2009). According to the Venture Impact website, a small number of high-tech start-ups financed by venture capitalists in 2003 created more than 10 million jobs and generated $1.8 trillion in sales, which represent 9.4 percent and 9.6 percent of total employment and sales in the United States, respectively (Shane, 2009). However, the minority or skeptical approach evaluates the entire picture of new firm formation. Accordingly, the stellar performance of a smaller number of gazelles is more than offset by the lackluster performance of a large number of typical small firms. For instance, recent research by the Global Entrepreneurship Monitor reports that necessity-driven entrepreneurship, which is mostly traditional small businesses, accounts for 36.5 percent, 25 percent, and 21.2 percent of all firm formation in Korea, Japan, and the United States, respectively (GEM, 2013). A relatively large proportion of necessity-driven new firms implies that the job and wealth contribution of small firms in Korea is likely smaller in comparison with the United States and Japan.

Such a dramatic difference in the two strands of new firms suggests that policy measures need to be tailored according to their different needs. Some scholars even argue that policy makers should stop subsidizing typical new firms and support only a subset of new firms with strong growth potential (Shane, 2009). However, this approach can also be as problematic as blinded support for all types of new firms, if an overwhelming number of new firms are typical traditional new firms with low growth potential. This is particularly the case in Korea, where a lion's share of new firms is indeed traditional self-employed businesses, mostly with the intention of substituting income.

Table 5.1 Industry classification for technology ventures

Manufacturing	Pharmaceuticals, medicinal chemicals and botanical products
	Electronic components, computer, radio, television and communication equipment and apparatuses
	Medical, precision and optical instruments, watches and clocks
	Electricity equipment
Information and communications	Computer programming, consultancy and related activities
	Information service activities
Professional, scientific and technical services	Architectural, engineering and other scientific technical services
	Professional, scientific and technical services, n.e.c.

Source: Bank of Korea and authors.

Spatial distribution of entrepreneurship in Korea

To analyze the spatial distribution of two strands of entrepreneurship in Korea, we first defined technology ventures and traditional small businesses based on industrial attributes. Industries associated with technology ventures rely mostly upon knowledge-intensive inputs, such as high-tech machinery and high-skilled workers, in their production activities. On the other hand, industries associated with traditional small businesses often draw their comparative advantage on cost-based replaceable inputs, such as low-skilled workers. Given a relatively low concentration of primary industries (i.e., agriculture and mining) in Korea, we focus the analysis only on the manufacturing and service sectors. The standard industry classification system in Korea categorizes industries into three levels. The mid-level classification, which this study draws upon, has 75 industries. We reviewed individual sectors in the mid-level classification and identified eight technology venture industries based on our judgment regarding the nature of their production process. A more detailed composition of technology venture industries is presented in Table 5.1. All other sectors in manufacturing and services are categorized as traditional small business industries. The number of employees in the industries associated with technology ventures is 2.9 million as of 2013, accounting for about 11.6 percent of the total employment as of 2013 in Korea.

The new firm formation data for this analysis were obtained from the Startup Statistics compiled by KISED. The data covers entrepreneurial activities in 15 metro areas and provinces from 2007 to 2013. Of the 1,638,738 new firms created across the country during this period, 1,598,956 are categorized as traditional small businesses. Technology venture industries produced only 2 percent of all entrepreneurial activities in Korea, which is much smaller than those industries' employment share (i.e., 6.1 percent). On the other hand, a lion's share of new firms was created in two sectors, "accommodation, food and beverage services" and "wholesale and retail trade," which account for slightly more than 50 percent of all new firm creation in Korea. This stark contrast can be partially explained by reviewing the motivation for starting a new business. According to a recent survey conducted by KISED (2013),[2] 82.2 percent of respondents answered that the most important driver of new firm creation was economic needs, whereas only 3 percent mentioned innovation and opportunity as a main driver for their entrepreneurship. This finding is well-aligned with an interesting survey conducted by the Global Entrepreneurship Monitor, which reports that Korean entrepreneurs show among the lowest level of perceived market opportunity across 70 economies around the world (GEM, 2013). When entrepreneurial activities are predominantly represented by only one type of new firms, policy measures need to be tailored accordingly since these two strands of firms are very different in nature. In particular, a tailored approach becomes even more important at the regional level because some regions have a unique composition of new firms.

Figures 5.1 and 5.2 shows firm formation activities in technology ventures and traditional small business sectors in "Gwangyuk" regions, which include 15 metro areas and provinces.[3] Several interesting patterns emerge. First, the formation of technology ventures is disproportionately concentrated in the capital region, which includes Seoul and Gyeonggi. The capital region is highly concentrated, accounting for 42 percent and 46 percent of the national total employment and establishment respectively. However, when it comes to high-tech entrepreneurial activities, the capital region claims 65 percent, which is much higher than its fair share considering employment and establishment concentration. In all other regions, the share of technology ventures is less than its total employment or establishment shares.

Second, the sectors, which we defined as technology venture industries (see Table 5.1), account for only 2 percent of all entrepreneurial activities. However, they represent 4 percent and 3 percent of entrepreneurial activities in Seoul and Gyeonggi respectively, which is much higher than the national average. On the other hand, in some regions such as Daejeon, Ulsan, Chungbuk, and Jeonnam, entrepreneurial activities in technology venture industries represent less than 1 percent. A relatively low profile of technology-oriented entrepreneurial activities in these regions can be partially explained. Ulsan is well known for the strong presence of the automobile cluster since Hyundai Motors has a strong grip on the regional economy. In addition, Jeonnam has a long tradition of agriculture and farming. Accordingly, it is not surprising that these two regions show relatively weak technology-oriented entrepreneurial activities. However, Daejeon and Chungbuk present a different story. Daejeon is a highly commended region for its successful development of a technology cluster, Daedeok Innopolis. Chungbuk is also known for its strong drive to develop a biotech cluster within the region. They are good candidates for hotbeds of technology-oriented entrepreneurial activities besides the capital region. However, the reality is markedly different from our expectations,

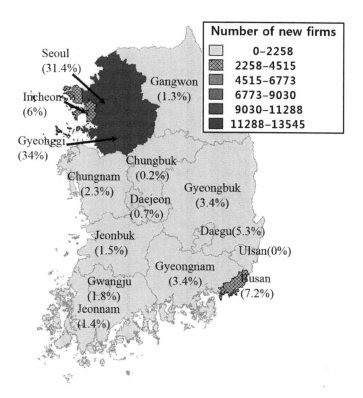

Figure 5.1 Geographic distribution of new firm formation in Korea – technology ventures
Source: KISED.

which implies that the development potential for these two regions as technology centers is not very rosy. In fact, all the high-tech entrepreneurs that we interviewed, answered that they prefer to locate in the capital region regardless of their current locations. This trend is particularly strong in IT-related activities, in which skilled labor is a key to success.

A popular belief is that technology ventures often serve as a key driver for long-term regional economic growth. However, the economic performance of the two regions (i.e., Seoul and Gyeonggi) with the highest level of technology-oriented entrepreneurial activities shows a mixed picture. Although these two regions are markedly different with respect to technology-oriented entrepreneurship from other regions, accounting for over 40 percent of the total new technology ventures in the country, their resemblance ends here. Table 5.2 shows that Gyeonggi was one of the fastest growing regions in the country, second to only Chungnam, during the 2007–2013 period, whereas Seoul has shown a lower than average economic performance. Considering that Chungnam's strong growth can be

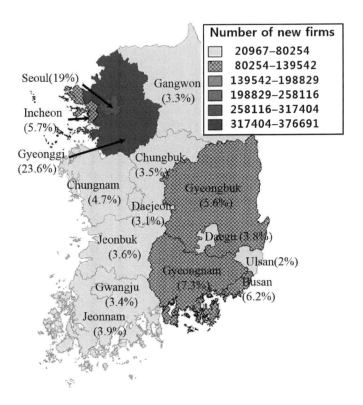

Figure 5.2 Geographic distribution of new firm formation in Korea – traditional small businesses

Source: KISED.

attributable to public sector relocation from the capital region, Gyeonggi can be categorized as the hottest high-tech region in the country. On the other hand, Seoul's growth rate and unemployment rate are the third lowest and the second highest in the country. Note that Seoul and Gyeonggi share locational advantages since they together constitute the capital region. This means that technology ventures, despite its significant presence, do not present strong growth potential in these regions. In addition, other fast growing regions in the country, such as Daejeon and Chungbuk, show the lowest level of technology venture activities. This circumstantial evidence suggests that technology ventures, at least in Korea, have not acted as a significant driver for the regional economy. As Shane (2009) eloquently warned, there seems to be no clear causal link between technology venture activities and regional economic growth in the case of Korea.

There are several reasons for the lack of regional economic influence of technology ventures in Korea. First, there are too few technology ventures.

Table 5.2 Regional economic performance in Korea

Region	Establishment (2013)	Employment (2013)	Annual growth (2007–2013)	Unemployment (2013)
Seoul	785,094	4,585,090	3.6%	3.5%
Busan	271,983	1,297,862	3.2%	2.9%
Daegu	195,717	849,631	3.9%	3.6%
Incheon	177,990	895,657	3.3%	3.8%
Gwangju	108,808	529,113	4.1%	2.1%
Daejeon	105,676	536,181	5.1%	2.8%
Ulsan	76,993	488,627	4.4%	3.0%
Gyeonggi	773,216	4,259,215	5.7%	2.7%
Gangwon	129,403	551,182	3.8%	3.1%
Chungbuk	115,611	591,509	5.6%	3.3%
Chungnam	145,998	777,843	7.4%	2.7%
Jeonbuk	139,656	624,407	5.0%	1.8%
Jeonnam	136,889	623,801	3.6%	2.5%
Gyeongbuk	205,734	1,004,067	4.3%	3.8%
Gyeongnam	248,913	1,275,688	4.7%	2.4%
Total/average	3,676,876	19,173,474	4.6%	3.0%

Source: Bank of Korea and authors.

Although the statistics published by the Korean government show that technology-based firm creations account for over 23 percent of all entrepreneurial activities (KISED, 2013), this can be misleading since technology ventures are defined very broadly in the government statistics, including most traditional manufacturing sectors, such as textile and paper products. If a more strict definition is imposed, as proposed in Table 5.1, technology ventures account for only 2.4 percent of all new firm formations in Korea during the study period. Even in Seoul and Gyeonggi, the hottest regions for technology ventures in Korea, the shares of technology ventures are only 3.9 percent and 3.5 percent of all new firm activities, respectively. Korea is often known for being an innovation-driven economy, mostly because of the strong presence of technology giants, such as Samsung and LG. Unlike Israel, where small technology ventures play a critical role in the country's innovation-led regional economies, a lion's share of technological innovations is driven by these global players rather than technology ventures.

Second, the market for technology, where entrepreneurs can capitalize on their achievements, is not fully developed in Korea. In 1998, Jeong Kim, a well-known Korean-American technology entrepreneur, sold his company, Yurie Systems, to Lucent for $1.1 billion, thus joining the ranks of well-known billionaires, such as Ross Perot and Chey Tae Won (owner of the SK Group, one of the big four Korean conglomerates). However, such a Cinderella story is unlikely to happen in Korea due to the absence of a well-established market for trading small technology firms. This is a critical drawback for developing a technology-driven entrepreneurial ecosystem in the regional economy, such as Silicon Valley. Accordingly, the major players at the forefront of innovation are mostly existing technology giants, such as Samsung and LG.

Third, the absence of a well-developed regional entrepreneurial ecosystem reduces the chance of finding a Cinderella story of a star venture, which often serves as a stimulant for many latent entrepreneurs to start their own businesses. This is a typical chicken-and-egg problem. The emergence of a star venture requires a nourishing environment for technology entrepreneurs. However, such an environment is more likely to be created when there are many active entrepreneurs in the market, which is often stimulated by the presence of star ventures.

In sum, technology ventures in Korea do not serve as a significant driving force for the regional economic performance. Their market presence is weak with respect to both number of firms and employment. Even technology hotbeds, such as Seoul and Gyeonggi, are no exceptions. This implies that the entrepreneurial policy at both the central and local levels should focus more on traditional small businesses rather than on technology ventures because they may have a bigger impact on the regional economy. However, a significant policy mismatch exists in reality.

Policy mismatches

Technology ventures are glamorous. Success stories of big name corporate stars, such as Steve Jobs and Bill Gates, attract talented individuals who want to be the next Jobs and Gates. It is not just technical talents who are attracted by the glamor of high-tech sectors. Policy makers also pay significant attention to promoting a regional environment conducive to technology-oriented entrepreneurial activities. Creating a new Silicon Valley has been a popular policy goal in many regions and countries, and promoting entrepreneurship is often located at the center of such efforts. Since the public sector lacks a concrete measure to evaluate performance in practice, the glamor of successful high-tech ventures provides a good means that can show the general public that policy makers in office are indeed working for the sake of the regional economy and its future. This approach is relevant and even desirable in an economy with a vibrant business and social environment for technology ventures, such as Silicon Valley. However, if the main players in the regional economy are predominantly large incumbents or traditional small firms, entrepreneurial policies in many regions and countries that are designed for promoting technology ventures are aimed at the wrong target. Accordingly, this mismatch may create significant inefficiency and inefficacy in entrepreneurial policy.

Korea is not an exception. We reviewed current entrepreneurial policies at both the central and local government levels. Detailed information about these policies are obtained from KISED. To review policies at the central government level, we focused on two agencies specialized in innovation and small firm

Table 5.3 Distribution of central government entrepreneurial policies and budgets

Focus areas	No. of technology venture targeted programs	No. of traditional small businesses programs	No. of mixed programs
Product development and commercialization	9 ($107,131,000)	2 ($2,100,000)	1 ($6,790,000)
Finance	8 ($18,164,000)	1 (-)	0 (-)
Incubator and mentoring (I/M)	12 ($72,711,000)	2 ($1,149,000)	10 ($30,834,000)
Education	3 ($13,200,000)	1 (-)	6 ($26,310,000)
Mixed focus	5 ($47,275,000)	0 (-)	0 (-)
Total	37 ($258,481,000)	5 ($3,249,000)	17 ($63,934,000)

Source: Korea Institute of Startup and Entrepreneurship Development.

support: the Ministry of Science, ICT and Future Planning and the Small and Medium Business Administration. In addition, we chose Seoul, Gyeonggi and Ulsan, Chungbuk from among the 15 regions in Korea since they show the highest and the lowest levels of technology-oriented entrepreneurial activities, respectively. The following interesting findings emerged from our policy review.

First, policies at the central government level focus largely on four areas: product development and commercialization, finance, incubation and mentoring, and education for potential entrepreneurs. In Table 5.3, product development and commercialization policies, which mostly aim at promoting technology start-ups with high growth potential, account for the largest share of entrepreneurial policy budget. This group of policies provides a wide range of support for bringing new technology ventures to the world market (e.g., prototype production and marketing). Table 5.3 also shows that incubation and mentoring policies for technology ventures rank first and second with respect to the number of programs and budget size, respectively. Although it is a relevant approach given the relatively small market size of Korea,[4] these policies are highly skewed toward technology ventures in comparison with other policies to promote traditional small businesses. Among the 59 programs that we reviewed, 37 programs (62.7 percent) are specifically designed to support technology ventures. Only five programs mainly target traditional small businesses. The difference in budget is even more striking. The total budget of technology venture targeted policies is $258.5 million, which is 86 times greater than that of traditional small business targeted policies.[5]

Particular attention needs to be paid to the dearth of specialized programs for traditional small businesses.[6] Although a lion's share of new businesses is created in traditional manufacturing, retail, and service sectors rather than in high-tech sectors, major government programs with sizable budgets are mostly designed to promote technology ventures. Such a policy mismatch can be partially attributable to the current administration's strong policy drive to

Table 5.4 Distribution of entrepreneurial policies and budgets in Seoul and Gyeonggi

Focus areas	No. of technology venture targeted programs	No. of traditional small businesses programs	No. of mixed programs
Product development and commercialization	0 (-)	0 (-)	0 (-)
Finance	0 (-)	0 (-)	1 (-)
Incubator and mentoring	6 ($2,044,000)	1 ($2,043,000)	0 (-)
Education	0 (-)	1 ($200,000)	0 (-)
Mixed focus	2 ($3,900,000)	0 (-)	1 ($14,640,000)
Total	8 ($5,944,000)	2 ($2,243,000)	2 ($14,640,000)

Source: Korea Institute of Startup and Entrepreneurship Development; the numbers in parentheses indicate budget.

promote a creative economy. However, it has been far from a success for the aforementioned several reasons.[7]

Entrepreneurial policies at the local level face a similar issue. We reviewed 12 major entrepreneurial policies currently in operation in the capital region (i.e., Seoul and Gyeonggi). Table 5.4 shows that eight of them specifically target technology entrepreneurship. In particular, Gyeonggi's policies are more highly skewed toward supporting technology ventures. Interestingly, the least entrepreneurial regions (i.e., Chungbuk and Ulsan), where the share of technology ventures is less than 1 percent, also have an almost similar policy distribution. Among nine entrepreneurial policies we reviewed in these regions, five of them aim to promote technology ventures (see Table 5.5). In addition, technology venture programs have much bigger budgets in comparison with traditional small business programs. This trend clearly suggests that most regions in Korea are overly committed to creating high-tech driven entrepreneurial economy when the reality says otherwise.

The policy mismatch can be found not only with respect to the distribution between technology ventures and traditional small businesses policies but also in its contents. Although the lion's share of existing entrepreneurial policies are tailored for promoting technology ventures, the voice of entrepreneurs in the field is not very positive. Most of the technology entrepreneurs we interviewed were critical about government policies at both the central and the local levels for several reasons.

First, the most significant criticism particularly focuses on the unrealistic nature of these policies. For instance, the five-year survival rate of new firms is less than 50 percent in Korea (KISED, 2013). It is even lower for need-based traditional small businesses in particular (less than 30 percent). However, according to our interviewees, most government programs at both the central and local levels target relatively more established new firms that

Table 5.5 Distribution of entrepreneurial policies and budgets in Ulsan and Chungbuk

Focus areas	No. of technology venture targeted programs	No. of traditional small businesses programs	No. of mixed programs
Product development and commercialization	2 ($30,000)	0 (-)	1 ($30,000)
Finance	1 (-)	0 (-)	0 (-)
Incubator and mentoring	3 ($245,000)	1 ($40,000)	1 ($140,000)
Education	0 (-)	0 (-)	1 ($12,000)
Mixed focus	0 (-)	0 (-)	0 (-)
Total	5 ($275,000)	1 ($40,000)	3 ($182,000)

Source: Korea Institute of Startup and Entrepreneurship Development; the numbers in parentheses indicate budget.

survived the first five years. These programs often require some sale records, although early start-ups usually need to secure funding to start manufacturing a prototype product.

Second, high-tech entrepreneurs were particularly discontented with the bureaucratic process. One of our interviewees confessed that it took over one year to receive the promised seed funding ($50,000) from the Small and Medium Business Administration, the agency responsible for supporting new entrepreneurs. The programs at the local government level are not greatly different. Delays were mostly attributable to the extensive paperwork that is often required to prevent fraud and embezzlement. Most new firms at this stage do not have any employees and are merely operated by their founders. These founders play multiple roles as CEO, manager, accountant, and marketer. The extensive paperwork required to receive government support creates an important barrier particularly for these early stage technology entrepreneurs. In fact, this environment sometimes leads to an adverse selection situation in which the government chooses new firms with sophisticated paperwork skills rather than innovative technology.

In the case of traditional small businesses, the situation is slightly different. A survey conducted by KISED (2013) reported that the difference in the share of business owners who have received some kind of support from government programs, at both the central and local levels, is particularly noticeable in the non-capital regions (technology ventures 14.7 percent vs. traditional small businesses 9.5 percent).[8] In addition, more traditional business owners in these regions expressed a need for financial support (technology ventures 35.9 percent vs. traditional small businesses 41.4 percent). Lastly, the satisfaction with government programs is much higher among traditional small business owners (65 percent) than among technology venture owners (41.2 percent). These survey results strongly indicate that traditional small businesses are in greater need of government support, particularly in non-capital regions.

When they receive needed support from government programs, they tend to be more satisfied. However, it turns out to be difficult for them to find support programs that fit their needs.

Similar to technology entrepreneurs, the small business owners we interviewed complained about the bureaucratic nature of central and local government policies. However, more importantly, they universally highlighted the information problem that they faced. According to our interviewees, entrepreneurs in the traditional sectors are less likely to secure support from a wide range of government programs. This is not because support programs for the traditional small businesses are less visible or skimpy but because new business owners are often unaware of what kind of support programs are available for traditional small businesses. This problem is well documented in the previously mentioned KISED survey (2013).

The information problem can be aggravated because several central government agencies and local governments often develop entrepreneurial policies in an unsystematic manner. Similar policies, often substantially overlapped, are scattered across different government agencies. The traditional small business owners that we interviewed responded that it was a Herculean task to find support programs that fit their needs. In addition, the overlapping programs with a similar nature require unnecessary and redundant paperwork, which results in significant administrative inefficiency.

Implications

Two beliefs about the Korean economy have gained wide currency: new firm formation is a driving force for regional development, and the economy is innovation driven. In combination, these two popular beliefs created a rosy picture of a high-tech entrepreneurial economy. However, the reality is markedly different from this misconception. We examined new firm formation activities in 15 regions in Korea. Although the capital region is far ahead of other regions with respect to technology entrepreneurship, technology ventures do not seem to play a critical role even in the capital region. Traditional small businesses are a more typical form of newly created firms, accounting for over 95 percent in all 15 regions. Clearly, Korea is not a high-tech entrepreneurial economy.

The revealed picture of entrepreneurship in Korea suggests that central and local governments need to pay more attention to traditional small businesses. In particular, the role of local governments in revitalizing the regional economy is critical. However, the distribution of entrepreneurial policy is highly skewed toward technology ventures, at both the central and local levels. Although the lion's share of new firm creation occurs in traditional business sectors, the policy support they receive is disproportionately less than technology firms. Undoubtedly, such a policy mismatch creates significant administrative inefficiency. This implies that policy redesign, particularly at the local level, may significantly improve the efficacy and efficiency of entrepreneurial policy. The aforementioned discussion and analysis of this study suggests the following implications for policy redesign.

First, both central and local governments need to shift their policy focus from high-tech ventures to traditional small businesses for practical reasons. It is more plausible to expect 100 successful small to medium firms in traditional sectors rather than one or two of the Samsung variety in high-tech sectors in the Korean context. Nurturing technology ventures aspiring to be the next Samsung is also a critically important strategy to upgrade the Korean economy, but one that should be treated as a separate and independent policy issue rather than an alternative to supporting traditional small businesses.

Second, local governments are better suited to supporting traditional small businesses because their owners are mostly locally based and tend to stay in the region. Therefore, regions need to strengthen their policy support for traditional small businesses that account for the majority of new job creation. Surveys and interviews with small business owners strongly indicate that policy makers should pay more attention to easing the financial burdens of the newcomers in the market. In addition, many traditional small businesses face difficulties in finding the right policy support because of insufficient information. Local government agencies need to revamp their efforts to inform entrepreneurs about the kinds of government support that are available.

Third, entrepreneurial policies, at both the central and local levels, need process innovations. Our interviewees universally described the unnecessary paperwork and tedious administrative procedures as an important barrier to receiving government support. Such procedural barriers prevent financial resources from being distributed to the right places at the right stages. This is particularly important for new entrepreneurs in traditional sectors because they are often relatively less educated in comparison with high-tech entrepreneurs and are in greater need of financial resources.

Lastly, unrealistic eligibility requirements need to be removed from entrepreneurial policies at both the central and local levels as they serve as another important barrier to securing government support. This issue is particular relevant for technology ventures that need funding support to produce prototype products whereas government support programs often require existing production records.

Acknowledgement

This work was supported by the National Research Foundation of Korea Grant funded by the Korean Government (NRF-2013-S1A3A2053959).

Notes

1 Although we considered the industrial and geographic distribution of our interviewees, these interviews may not reflect the views of most entrepreneurs.
2 This is a nationwide survey for entrepreneurship in Korea. The survey covers all manufacturing and service industries in 16 "Gwangyuk" regions and has some 6,000 respondents. The subject firms are less than seven years old.

3 Korea has a two-layered local government system. The first layer is called "Gwangyuk" and includes 16 medium to large scale metros and provinces. This study uses 15 "Gwangyuk" regions for analysis. The excluded region (Jeju) is an island that has somewhat different economic characteristics in comparison with the other 15 regions. The second layer is called "Gicho" and includes 234 small scale cities and regions. These two layers of local governments have a hierarchical structure. The "Gwangyuk" regions consist of multiple "Gicho" regions.

4 In the highly uncertain world of technology ventures, a large market is a necessary condition to recover from a series of failures with one success.

5 Information on program budget is somewhat limited. Among 82 policies that we reviewed at both the central and local levels, 24 policies do not provide detailed program budget information.

6 There are of course government programs designed to support traditional small businesses. But, they are often scattered across many central and local government agencies, and the program budgets are relatively small in comparison with the policies we reviewed.

7 In fact, the aforementioned problems, with respect to technology ventures in Korea, require more structural reform than a symptomatic approach at the policy or program levels.

8 We analyzed the raw data from the survey according to our categorizations of technology ventures and traditional small businesses.

References

Acs, Z. J., Audretsch, D. B., & Carlsson, B. (2003). The knowledge filter and entrepreneurship in endogenous growth. Paper presented at the North American Regional Science Council, Philadelphia.

Aghion, P., Blundell, R., Griffith, R., Howitt, P., & Prantl, S. (2004). Entry and productivity growth: Evidence from microlevel panel data. *Journal of European Economic Association*, 2, 265–276.

Baptista, R., & Preto, M. T. (2011). New firm formation and employment growth: Regional and business dynamics. *Small Business Economics*, 36, 419–442.

Birch, D. (1981). Who creates jobs? *Public Interest*, 65, 4–14.

Birch, D. (1987). *Job Creation in America*. New York: Free Press.

Broersma, L., & Gautier, P. (1997). Job creation and job destruction by small firms: An empirical investigation for the Dutch manufacturing sector. *Small Business Economics*, 9, 211–224.

Davis, S., Haltiwanger, J., & Schuh, S. (1996). Small business and job creation: Dissecting the myth and reassessing the facts. *Small Business Economics*, 8, 297–315.

Davisson, P., Lindmark, L., & Olofsson, C. (1994). New firm formation and regional development in Sweden. *Regional Studies*, 28, 395–410.

De Kok, J., De Wit, G., & Suddle, K. (2006). SMEs as job engine of the Dutch private economy: A size class decomposition of employment changes for different sectors of the Dutch economy. EIM Research Report H200601.

Disney, R., Haskel, J., & Heden, Y. (2003). Entry, exit and establishment survival in UK manufacturing. *Journal of Industrial Economics*, 51, 91–112.

Feldman, M. P. (1994). *The Geography of Innovation*. Dordrecht, Boston and London: Kluwer Academic Publishers.

Fritsch, M., & Mueller, P. (2004). Effects of new business formation on regional development over time. *Regional Studies*, 38, 961–975.

GEM (2013). *Global Entrepreneurship Monitor 2013 Global Report*. Global Entrepreneurship Monitor.

Haltiwanger, J., Lane, J., & Speltzer, J. (1999). Productivity differences across employers: The role of employer size, age, and human capital. *American Economic Review Papers and Proceedings*, 89, 94–98.

Henrekson, M., & Johansson, D. (2009). Gazelles as job creators: A survey and interpretatin of the evidence. *Small Business Economics*, 35, 227–244.

KISED (2013). *Startup Statistics*. Deajeon: Korea Institute of Startup and Entrepreneurship Development.

Mueller, P., Van Stel, A., & Storey, D. J. (2008). The effects of new firm formation on regional development over time: The case of Great Britain. *Small Business Economics*, 30, 59–71.

Picot, G., & Dupuy, R. (1998). Job creation by company size class: The magnitude, concentration and persistence of job gains and losses in Canada. *Small Business Economics*, 10, 117–139.

Shane, S. (2001). Technological opportunities and new firm creation. *Management Science*, 47, 205–220.

Shane, S. (2009). Why encouraging more people to become entrepreneurs is bad public policy. *Small Business Economics*, 33, 141–149.

Van Praag, M., & Versloot, P. (2007). What is the value of entrepreneurship? A review of recent research. *Small Business Economics*, 29, 351–382.

Van Stel, A., & Suddle, K. (2008). The impact of new firm formation on regional development in the Netherlands. *Small Business Economics*, 30, 31–47.

Verhoeven, W., Gibcus, P., & De Jong Hart, P. (2005). *Bedrijvendynamiek in Nederland: Goed of Slecht?* Zoetermeer: EIM.

6 The development of entrepreneurship in China

A geographical and institutional perspective

Canfei He, Qi Guo, and Shengjun Zhu

Introduction

For much of the second half of the twentieth century, China and entrepreneurship have been rarely mentioned together. Even though the Chinese diaspora had a long, phenomenal history of entrepreneurship at the global scale, entrepreneurship had been largely suppressed and stifled for decades in mainland China since the Communist Party came into power in 1949. However, since the initiation of China's Reform and Opening-Up policies, China has undergone dramatic economic growth and has experienced three fundamental transformations: (1) from a centrally planned to an increasingly decentralized political and economic system; (2) from a state-owned, collective economy to one with growing level of private ownership and market orientation; and (3) from a partially closed economy to one oriented toward export markets (He et al., 2008; Wei, 2001). The transition process has brought about the liberalization of prices, markets, and trade, and the privatization of the state-owned sectors. Decentralization has granted local governments authority to intervene in economic development. Moreover, globalization has brought international competition into local development. Firms face challenges from market competition and institutional uncertainties. Driven by this export-oriented strategy, China's average annual GDP growth rate was around 9.8 percent; exports increased by 12.4 percent annually in the 1990s and by more than 20 percent a year in the 2000s before the outbreak of the global financial crisis in 2008 (Zhu & He, 2013). As marketization, privatization, and globalization proceed in China, the power of entrepreneurship flourished in response to the internal economic and political reform.

Over the years, an increasing number of academic professional studies have focused on Chinese entrepreneurship from the 1990s, by paying special attention to social networks and social capital connected with family ties (Batjargal & Liu, 2004; Bates, 1997), entrepreneurial culture and values (Tan & Chow, 2009), and co-evolution of entrepreneurship strategy and entrepreneurial environment during the economic transition (Lewin et al., 1999). This chapter builds on these existing studies and seeks to examine the development of entrepreneurship from a geographical and institutional perspective.

Entrepreneurship in China: a brief history

In the first half of the twentieth century, central and local administration's underestimation, if not neglect, of entrepreneurial ventures and widespread warfare and upheaval for decades all curbed the development of entrepreneurship in China (Landes, 1999). After the Communist Party took over, China was brought by Mao into a socialist system that collectivized agriculture and nationalized industry, leaving little room for private sector activity. Following the Soviet central planning model, resources and factor inputs were disproportionately allocated by the state toward the development of heavy industry (Naughton, 1996). Despite this, entrepreneurship was not completely eliminated and remained at a small scale, particularly in the form of the black market and underground economy. Even though such activity was largely unproductive at this time, actors in these markets were able to profit by taking advantage of economic shortages in the 1960s and 1970s.

After the Cultural Revolution, with the economy in very poor shape, Deng Xiaoping, China's new leader, launched the Four Modernizations in 1978 and the Reform and Opening-Up policies. The de-collectivization of agriculture generated rural unemployment and an impetus for the rapid development of town and village enterprises (TVEs), which made up 20 percent of China's gross output by 1990 (Liao & Sohmen, 2001). Although the managers of TVEs were not completely modern entrepreneurs as the enterprises were collectively owned by local authorities, they did have some characteristics of entrepreneurs as they not only were in charge of financing, producing, marketing, and distributing, but also acted in response to market prices and chased for profits. However, real private enterprises were not formally permitted to exist until 1988 – ten years after the initiation of the Reform and Opening-Up policies; the institutional environment (e.g., property rights protection and contract enforcement) still did not favor entrepreneurial ventures (Lu & Tao, 2010). Constraints on entrepreneurship were further lifted by Deng's South China tour in 1992 and by the privatization of a large number of state-owned enterprises in 1998. To stimulate economic growth and regional development, the state increasingly encouraged the formation of private enterprises, liberalized foreign trade and investment, relaxed state control over the economy, and heavily invested in labor training and education.

Entrepreneurship in China: an institutional perspective

North's (1990) seminal work has pointed out the impact of a country's political, economic, and social institutions on economic performance, firm operations, and entrepreneurial activities. Unlike entrepreneurship in developed economies that are largely determined by the personal attributes of entrepreneurs and the market economy, the institutional context in which Chinese entrepreneurs operate is institutionally dynamic, which complicates the entrepreneurial process (He et al., 2008; Wei, 2001). Specifically, China's institutional context has been constantly shaped by a triple process of marketization, globalization, and decentralization triggered by the Reform and Opening-Up policies since the late 1970s.

Marketization

Since the late 1970s, China's economic system has been transforming from a command economy in which the state was the planner of the economy and entrepreneurship was largely prohibited, toward a more market-oriented one, as privatization and market competition was progressively introduced into the economy (McMillan & Naughton, 1992). Since the early 1980s, the restrictions on business licensing have been gradually lifted and entrepreneurship has been encouraged in some regions and industries (Chang & MacMillan, 1991). Non-state capital and privately owned firms started to exert an increasingly important effect in economic development (He et al., 2008). Marketization generated freer and fairer competition environment and provided production factors, inputs, and demand for enterprises in a much more efficient way. In such a market-oriented economy, firms became more motivated to utilize comparative advantages and more likely to locate close to specialized suppliers and customers, resulting in industrial clustering and agglomeration. The favorable market environment as well as industrial clustering allowed new firms to reduce not only logistics costs, but also costs of searching for matched suppliers and customers. As a consequence, economically liberalized regions were more likely to provide entrepreneurs with a stable and transparent business environment, characterized by low institutional barriers and bureaucratic costs as well as efficient market competition. In addition, the large presence of privately owned firms could create an entrepreneurial climate, where entrepreneurial failure was tolerated easily and a large amount of knowledge on enterprise management and market opportunities could spill over from successful precursors to start-ups. In a word, marketization plays a positive and crucial role in enhancing entrepreneurship.

China's marketization reform has unleashed economic vitality, but its effectiveness is quite uneven across regions and industries in China. The coastal region is the first to benefit from the marketization process, which then progressively spreads to inland China (Naughton, 1996). The impacts of marketization also vary across industries. In the early stage of reform, the state first encouraged private enterprises to enter into light and labor-intensive industries, and then lowered the state-owned proportion in other manufacturing industries (Wei, 2001). Until now, state-owned capital is still dominant in certain industries, such as tobacco processing and processing of petroleum, coking, processing of nuclear fuel. The variation of marketization may be one of the important reasons behind the spatial and sectoral variations in entrepreneurship (He et al., 2008). One of the goals of this study is to explore if entrepreneurship appears to be influenced by marketization in different regions and industries.

Globalization

As China's Reform policies pushed forward the process of marketization in the economy, the Opening-Up policies globalized China's economy and allowed Chinese enterprises to access financial capital, more advanced technologies, knowledge and management skills, and high quality input factors on the international market (He et al., 2008). Enterprises benefited from globalization as they learned from foreign direct investment (FDI) that started to swarm into China in the 1980s on the one hand, and were increasingly encouraged to establish extra-regional linkages with firms in the North, on the other.

However, the impact of FDI over domestic entrepreneurship is theoretically and empirically inconclusive. On the one hand, FDI can foster domestic entrepreneurship through the spillover of technology and managerial experience (Blomström & Kokko, 1998), demonstration effects (Caves, 1996), and through production linkages, especially the provision of input and demand (Javorcik, 2004). On the other hand, FDI may crowd out domestic entrepreneurship by raising barriers to entry. Foreign-owned firms also compete with domestic ones for production factors and market (Aitken & Harrison, 1999; Haddad & Harrison, 1993). In some cases, local governments' fixation on attracting large transnational corporations to boost the local economy may lead to an institutional context that disadvantages domestic firms.

Empirically, some studies report positive spillover effects from foreign-owned firms to domestic ones (Cheung & Lin, 2004; Liu, 2008; Liu et al., 2002). It is argued that foreign-owned firms have had significant positive vertical effects on domestic enterprises (Lin et al., 2009; Liu, 2008). However, other studies disagree and find FDI has no spillover effects or even negative effects on domestic firms, especially in transition economies (e.g., Chen et al., 1995; Djankov & Hoekman, 2000; Konings, 2001; Wei, 2002). For example, Lin et al. (2009) have demonstrated that Hong Kong-, Macao-, and Taiwan-owned firms have negative horizontals spillover effects in China, whereas FDIs from Europe and North America have more positive effects. Specifically, FDIs are likely to have negative impacts over domestic firms in the same industry and such impacts are particularly remarkable in low-tech industries (Jeon et al., 2013). De Backer and Sleuwaegen's (2003) research provides evidence for the crowding-out effects of FDI on domestic entrepreneurship on both product and labor markets. Ayyagari and Kosová (2010) find that service industries benefit from FDI spillovers, while manufacturing industries do not see significant positive entry spillovers. Overall, the impact of FDI on entrepreneurship in China is unclear and still demands empirical investigations.

Decentralization

With the deepening of economic reform, decentralization from the central to the local has gradually granted local governments more autonomy, which started to have a primary responsibility for economic development in their respective jurisdictions. The increasing decentralization affects domestic entrepreneurship in two directions. On the one hand, fiscal decentralization has triggered fierce inter-regional competition. Local governments take a variety of measures to protect industries with high tax-profit margins and state-owned proportion under their administrations since these industries are the base of political power as well as the source of private benefits and fiscal revenues (Bai et al., 2004; Zhao & Zhang, 1999). Some favorable policies, in the form of tax credits, financial and technological aids, and cheap land, have been implemented to protect these industries, in order to increase profit margins and reduce risks in these industries. Although the protection of particular industries is justified as a way to improve fiscal revenues and to promote economic growth (Development Research Center of the State Council, 2004), it does not help the development of private sectors and domestic entrepreneurship, which is more likely to be incubated in a fair, competitive, and market-oriented environment. Overall, we expect that the level of entrepreneurship is lower in protected industries and regions in China.

On the other hand, decentralization has given local governments more power to intervene in local and regional economic development. Pro-growth local governments that pursue long-term economic growth often provide a variety of subsidies for start-ups or launch some programs to support entrepreneurship. These subsidies not only encourage potential entrepreneurs to start new businesses, but also allow new firms to sustain more easily. The supportive programs are found to significantly influence the development of entrepreneurship. For instance, Keuschnigg and Nielsen (2004) report that governmental subsidies are clearly effective in encouraging firm entry. Crépon and Duguet's (2003) findings also support the positive effect of subsidies on the survival of new firms. Therefore, we expect that entrepreneurship is more likely to occur in industries and regions with more subsidies.

Overall, entrepreneurs in China have to deal with the opportunities and challenges generated by the triple process of globalization, marketization, and decentralization. This study seeks to examine the development of China's entrepreneurship from a geographical and institutional perspective, by paying special attention to the ways in which firm dynamics in China have been constantly co-shaped by globalization, marketization, and decentralization.

Who are China's entrepreneurs and how to measure entrepreneurship in China?

There are three types of entrepreneurs in the context of China (Liao & Sohmen, 2001). The first type emerged in the 1980s as street vendors doing small-scale activities in retail and services. This group became self-employed as most of them were uneducated, illegal migrants and therefore excluded from the state system. Some succeeded, particularly after the reform, and became real entrepreneurs. The second type was highly educated and used to work in the state system (e.g., state-owned enterprises). After leaving the state system and starting their own private businesses, they still maintained strong connections with the public sector. The third type refers to the foreign-educated returnees, most of whom became quite active more recently, particularly in the Internet sector.

Individual or company survey data has been widely used in studies on entrepreneurship in China. Although it provides important information on entrepreneurial traits, the sample is often limited. In this chapter, one database on new firm formation and firm-specific economic and financial variables is used to measure and investigate entrepreneurship in China: China's Annual Survey of Industrial Firms (ASIF) (1998–2008). The ASIF is administered by the National Bureau of Statistics of China and covers all Chinese industrial state-owned enterprises and non-state-owned enterprises with annual sales of 5 million RMB or more. Year 2004 is China's economic census year, and the 2004 database includes many firms that have been omitted in non-census years. As a result, statistics in 2004 are quite erratic and not taken into account in this research. The database provides firm-level data on firm structure and operation, including firm official IDs, sector types, location, capital structure, total profits, total shipments, exported shipments, intermediary inputs, asset value, inventory, employment, sales value, type of investment, output, value added, R&D expenses, education and training of staff, and wages, social insurance, and benefits paid. One weakness of this database is that some small firms may be excluded. Following Glaeser and Kerr (2009) and Delgado et al. (2010), we measure entrepreneurship as the formation of private-owned start-ups, calculated as the number of start-ups and the employment in these start-ups during their starting year.

Temporal and spatial pattern of entrepreneurship in China

Table 6.1 shows the rapid rise of entrepreneurship in China during 1998–2007. Specifically, the number of private-owned start-ups soared by 390 percent from 1,507 in 1998 to 7,397 in 2007. Likewise, the employment of privately owned start-ups rose dramatically by 248 percent from 262,678 to 654,453 during this period. The share of privately owned start-ups of all new start-ups has increased rapidly during 1998–2007, suggesting entrepreneurship plays an increasingly important role in China's economy. In 2003, privately owned start-ups started to dominate as they accounted for more than 50 percent of the total new start-ups. Furthermore, the development of entrepreneurship in China also mirrors the broader industrial restructuring taking place in China's economy. On the one hand, the average firm size of privately owned start-ups (see average employment of privately owned start-ups in Table 6.1) decreased from 174 in 1998 to 88 in 2007, indicating that the entry barrier for new start-ups has been gradually lowered due to the increasingly favorable market environments, such as more flexible capital, labor, policies, and supplier–customer

Table 6.1 Temporal development of entrepreneurship in China (1998–2007)

Year	Number of private-owned start-ups	Employment of private-owned start-ups	Share of private-owned start-ups	Share of private-owned exporting firms in all private-owned start-ups	Average employment of private-owned start-ups
1998	1,507	262,678	26.6%	15.3%	174
1999	855	151,963	24.5%	15.3%	178
2000	963	161,727	34.3%	14.0%	168
2001	2,333	407,868	44.6%	16.5%	175
2002	1,709	241,300	48.3%	11.4%	141
2003	3,502	484,719	51.1%	11.2%	138
2005	5,222	564,461	51.0%	7.5%	108
2006	5,771	533,767	56.7%	7.1%	92
2007	7,397	654,453	55.3%	6.2%	88

Source: China's Annual Survey of Industrial Firms (ASIF).

linkages. On the other hand, the share of privately owned start-ups with an export orientation in all privately owned start-ups has decreased from 15.3 percent to 6.2 percent from 1998–2007, suggesting a new round of transition from export-oriented industrialization to one that is increasingly based on competitiveness of domestic entrepreneurship.

Figures 6.1 and 6.2 show the spatial distribution of entrepreneurship at the city level in 1998 and 2007, respectively, measured as the number of start-ups. A significant geographical variation of entrepreneurship can be observed in Figures 6.1 and 6.2. First, there is a high level of entrepreneurship in China's wealthy coastal region, where the reform first started, and in some regional centers (e.g., Chongqing) along China's Yangtze River. Central China also has a large number of new firms during this time period. In contrast, China's vast western regions have relatively underperformed for decades. Second, geographical variation also exists within China's coastal region. Cities in a number of clusters have stood out from the crowd in terms of the development of entrepreneurship, such as cities in the Yangtze River Delta, Pearl River Delta, Greater Beijing Region, and the Shandong Peninsula. Some peripheral cities in the coastal region also lag behind, particularly in North Jiangsu, East Guangdong, and Northeast Fujian.

Table 6.2 shows the spatio-temporal change of entrepreneurship during 1998 and 2007 at the provincial level. As shown in Figures 6.1, 6.2, and Table 6.2, some traditionally developed regions, for example, in the Yangzi River Delta and Pearl River Delta, have had a high level of entrepreneurship throughout the study period. Meanwhile, regions in Central China, such as Jilin, Hunan, Jiangxi, and Anhui province, have experienced a rapid growth of entrepreneurship. The most striking example in Figures 6.1, 6.2, and

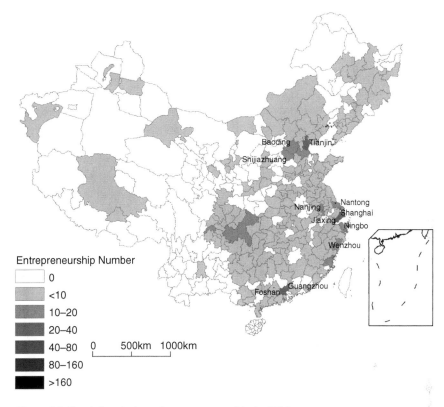

Figure 6.1 Spatial variation of entrepreneurship in 1998
Source: China's Annual Survey of Industrial Firms (ASIF).

Table 6.2 is that entrepreneurship in the Shandong province has increased dramatically, accounting for around 25 percent of the total in China in 2007. However, in the largest Chinese cities, Beijing and Shanghai, the number of start-ups has declined during this period. Since this database only includes firms in the secondary industry, this decline of entrepreneurship may indicate a transition from manufacturing to service industries in Beijing and Shanghai in the 2000s.

Spatial concentration of entrepreneurship

In order to further investigate the spatial pattern of entrepreneurship, an entropy index is used to quantify the geographical concentration of entrepreneurship and decompose the overall entropy at the city level into two parts – entropy between provinces (H_b) and entropy within province (H_g). Following Hexter and Snow (1970), we calculate the formulas below.

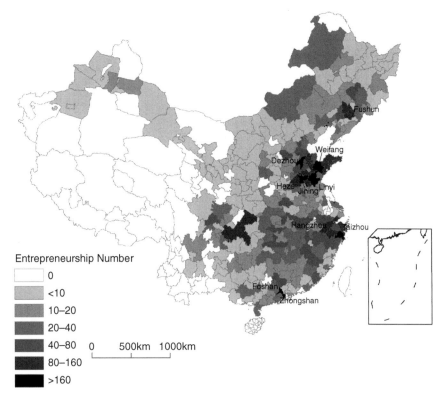

Figure 6.2 Spatial variation of entrepreneurship in 2007
Source: China's Annual Survey of Industrial Firms (ASIF).

$$H = \sum_{i=1}^{n} p_i \ln\left(\frac{1}{p_i}\right) = H_b + \sum_{g=1}^{G} p_g + H_g$$

$$H_b = \sum_{g=1}^{31} p_g \ln\left(\frac{1}{p_g}\right) p_g = \sum_{i \in s_g} P_i, \ H_g = \sum_{i \in s_g} \frac{P_i}{p_g} \ln\left(\frac{1}{P_i/P_g}\right) g = 1, \dots 31$$

Where p_i is the share of city i's privately owned start-ups of the national total; P_g is the share of province g's privately owned start-ups of the national total. Thus, H is the overall entropy at the city level, which is the sum of H_g and H_b. The former is the entropy within provinces while the latter is the entropy between provinces. $i \in S_g$ means that city i belongs to province g. A high entropy index suggests that entrepreneurship is more geographically dispersed.

Table 6.2 Spatio-temporal change of entrepreneurship at the provincial level (1998 and 2007)

Province	1998	2007	Annual average growth rate (%)	Province	1998	2007	Annual average growth rate (%)
Jilin	3	167	155.98	Shanxi	20	68	110.22
Jiangxi	16	284	136.77	Xinjiang	8	27	110.09
Shandong	120	1922	135.12	Heilongjiang	26	77	107.77
Hunan	18	288	135.11	Ningxia	3	8	105.84
Anhui	31	367	130.32	Hebei	105	274	105.43
Guizhou	4	45	129.51	Gansu	9	22	104.17
Guangdong	91	745	124.50	Henan	127	237	98.42
Hubei	43	335	123.72	Qinghai	2	3	92.59
Fujian	36	223	120.09	Shaanxi	25	34	89.27
Neimenggu	19	116	119.86	Tianjin	25	30	83.63
Liaoning	79	470	119.45	Hainan	2	2	0.00
Chongqing	19	108	118.72	Shanghai	40	20	−92.59
Guangxi	17	92	117.93	Jiangsu	265	108	−94.35
Zhejiang	248	992	112.98	Yunnan	6	2	−95.59
Sichuan	82	326	112.88	Beijing	18	5	−96.45

Source: China's Annual Survey of Industrial Firms (ASIF).

Table 6.3 shows the temporal change of the spatial concentration of entrepreneurship during 1998–2007. First, after an initial increase from 1998–2003, the overall entropy at the city level started to decrease in 2005, indicating an increasing geographic concentration of entrepreneurship, particularly in the second half of this time period. Second, even though in the late 1990s, the entropy between provinces increased and that within provinces decreased, this comparison reversed in the 2000s. The rising intra-provincial entropy and declining inter-provincial entropy in the 2000s indicate that entrepreneurship became increasingly concentrated at the provincial level while within province it started to disperse.

China's decentralization may contribute to the rising intra-provincial entropy and declining inter-provincial entropy in the 2000s in two ways. First, entrepreneurship refers to a process where entrepreneurs search for market opportunities and seek to combine resources to establish new businesses in an uncertain and in some cases risky environment (Dollinger, 2002). Entrepreneurs have to face complex and uncertain entrepreneurial environments even in developed economies where markets are more mature and transparent, let alone in China which is transitioning rapidly from a planned economy to a market economy. The frequent changes of institutional environments and policies bring in additional risks and insecurity to entrepreneurs. As a result, entrepreneurs are more likely to start their

Table 6.3 Temporal change of the spatial pattern of entrepreneurship (1998–2007)

Year	Overall entropy	Entropy between provinces	Entropy within province	Share of entropy between provinces	Share of entropy within provinces
1998	4.80	2.76	2.04	57.6%	42.4%
1999	4.83	2.99	1.85	61.8%	38.2%
2000	4.86	3.04	1.82	62.5%	37.5%
2001	4.92	2.87	2.05	58.4%	41.6%
2002	4.95	2.89	2.06	58.3%	41.7%
2003	5.00	2.83	2.17	56.7%	43.3%
2005	4.89	2.59	2.30	53.7%	46.3%
2006	4.64	2.38	2.26	51.2%	48.8%
2007	4.88	2.59	2.29	53.3%	46.7%

Source: China's Annual Survey of Industrial Firms (ASIF).

new business in the environments that they are familiar with. The entrepreneurial policies thus play an important role in the development of entrepreneurship. These policies are generally similar within the same province, but significantly different across provinces, as China's decentralization mainly takes place at the provincial level. In other words, decentralization has granted provincial governments, rather than city governments, greater authority to intervene in local economic development within their respective jurisdictions. Local governments tend to provide subsidies, loans, rebates of value added tax, and low-price land to indigenous enterprises, and enterprises need to pay extra fees when they enter other regions. Familiarity with specific industrial policies and institutional environments is important for entrepreneurs to successfully start their new businesses. As a result, the fact that the inter-provincial variation of these policies and institutions is large and intra-provincial variation is rather small results in a rising inter-provincial variation in terms of entrepreneurship and a declining intra-provincial variation.

Second, some low-end, labor-intensive industries are facing the challenges of rising production costs (e.g., labor and land cost) in China's developed coastal region, particularly since the early 2000s. They have also become increasingly unwelcome by local governments in the coastal provinces, as the latter seek to upgrade from low-end, labor-intensive production toward high-value, high-end, and high-tech industries. Enterprises in such industries are encouraged or forced by local governments in the coastal region to relocate to labor-intensive, low-end production in low-cost sites within inland China (or in some cases outside China, such as Southeast Asian countries) while retaining the high-end function (such as R&D and headquarters) in the coastal region. In doing so, local governments in the coastal region are able to sustain economic growth and to promote industrial upgrading. On the other hand, local governments in inland China consider this relocation an opportunity to attract more investment and boost the local economy, and therefore issue favorable policies to attract relocating firms from the coastal region (e.g., Foxconn's relocation from Guangdong to Henan). Decentralization and the subsequent inter-provincial competition on economic development have triggered a complicated process of industrial upgrading in the coastal region and industrial relocation to inland China, which also contributes to the increasing inter-provincial variation of entrepreneurship, particularly between coastal provinces and provinces in inland China.

A typology of entrepreneurial cities in a marketized, globalized, and decentralized China

To examine the impact of China's economic transition on the spatial pattern of entrepreneurship, this section classifies cities into different categories based on the degree to which cities are influenced by marketization, globalization, and decentralization, and then compares entrepreneurship in different types of cities. Figure 6.3 shows the temporal change of entrepreneurship in different types of cities during 1998–2007. Three-hundred-and-forty-three cities are investigated in this study. The abnormally high value in 2004 in all four figures may be due to the fact that Year 2004 is China's economic census year and the 2004 database includes many firms that have been omitted in non-census years. Overall, entrepreneurship has developed rapidly in all types of cities during this time period.

Market-oriented cities are defined as ones ranking in the top 50 percent in terms of the share of private capital in the total, which contain 172 cities. In line with our expectation, the level of entrepreneurship in market-oriented cities is much higher than that in less market-oriented cities during 1998–2007 (see Figure 6.3a), suggesting that marketization has provided a more favorable environment for entrepreneurs and privately owned start-ups in China.

If a city exports more than 6 percent of its gross industrial output or more than 4 percent of the total capital in this city is made up by foreign-owned capital, it will be considered a globalized city. Based on such criteria, there are 187 globalized cities, most of which are located in China's coastal region. Figure 6.3b shows a positive relationship between globalization and entrepreneurship in Chinese cities; the globalized cities have a higher level of entrepreneurship than other cities. A possible explanation for this relationship is that globalization can foster entrepreneurship at the city level because domestic firms are able to access advanced technologies and know-how by co-locating with foreign-owned firms or by participating in the global economy and establishing linkages with actors in the North.

Regional decentralization has granted local governments more authority to intervene in local economic development. Some local governments tend

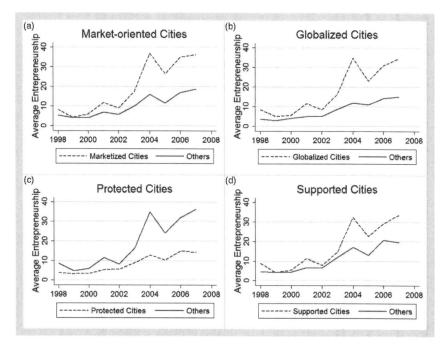

Figure 6.3 Entrepreneurship in different types of cities during 1998–2007
Source: China's Annual Survey of Industrial Firms (ASIF).

protect state-owned enterprises and highly profitable industries to promote fiscal revenues and economic growth, while others that pay more attention to long-term economic development launch programs to nurture entrepreneurship and focus less on protecting certain industries. We define protected cities as those with state capital over 50 percent of the total or with over 10 percent profit-tax rates. In doing so, we end up with 165 cities. Figure 6.3c shows that highly protected cities have a lower level of entrepreneurship, implying that conservative policies that aim to protect certain industries may not be conducive for entrepreneurship.

Supported cities are those ranking in the top 50 percent in terms of the share of subsidies provided for privately owned firms in the total, which contain 172 cities. Figure 6.3d shows that supported cities have a higher level of entrepreneurship, suggesting that subsidies can attract potential entrepreneurs to start their own business and facilitate the survival of start-ups. Since entrepreneurship has been justified as a key driver of sustainable economic development (Djankov & Murrell, 2002; Schumpeter, 1934), the positive effect of subsidies on entrepreneurship indicates the importance of subsidies in creating a nurturing environment for start-ups and sustaining long-term economic development.

Conclusion

Entrepreneurship has been critical for China's economic development, particularly after the economic reform. This chapter explores the temporal and spatial patterns of entrepreneurship in China from an institutional perspective, by paying special attention to the ways in which China's entrepreneurship has been shaped by a triple process of marketization, globalization, and decentralization. Using this analytical framework, we find that global forces, market forces, and government intervention all play a critical role in the development of entrepreneurship in post-Reform China.

Overall, entrepreneurship in China has developed dramatically during 1998 and 2007, the time period for this study. Alongside the rapid rise of entrepreneurship, particularly after the economic reform, is a more complicated process of industrial and geographical restructuring of China's entrepreneurship likely driven by marketization, globalization, and decentralization. First, there are significant spatial variations of entrepreneurial activity in China. The coastal region maintains the highest level of entrepreneurship during this period. The level of entrepreneurship in the central region is relatively lower than the coastal region, but has experienced rapid growth. The western region presents the lowest level of entrepreneurship. Second, the spatial distribution of entrepreneurship is increasingly concentrated. Furthermore, inter-provincial variation of entrepreneurship has increased whereas intra-provincial variation has been declining, due in large part to the decentralization strategy launched by the state after the reform.

We further compare entrepreneurship in different types of cities, which are classified as globalized ones, market-oriented ones, protected ones, and ones with high subsidies and others. In line with our expectation and theories, globalization and marketization at the city level likely provide a more favorable environment for entrepreneurship. Protection policies appear to have a negative relationship while subsidies have a positive relationship with entrepreneurship in Chinese cities.

Entrepreneurship has always been an important topic, especially in transition economies such as China. We hope that the institutional framework based on the recent economic transition can provide a useful lens to understand the development of Chinese entrepreneurship, as well as its temporal and geographical evolution. This study, however, only investigates its spatial pattern by using highly aggregated data. Future research can be directed to explore the industrial variations of entrepreneurship in China at a more disaggregated level. The dataset we use in this study ends in 2008, therefore more up-to-date data will allow for a more recent examination of entrepreneurship in China, particularly as regards the impact of the 2008 global financial crisis on China's level of entrepreneurial activity. After the financial crisis, two types of industrial restructuring unfolded in China. On the one hand, Chinese enterprises started to re-orient their production from export to domestic markets, due mainly to the slackening global demand for China's exports particularly after the global financial crisis. On the other hand, as Chinese enterprises upgrade, they increasingly seek to enter the international market and to sell branded products to customers overseas. We leave the articulation between these two types of industrial restructuring and China's entrepreneurship for future research. Finally, case studies and qualitative researches based on survey and firm interviews would also serve as a complement to the present study.

Acknowledgement

This research is funded by the National Science Foundation of China (No. 41425001 and No. 41271130).

References

Aitken, B. J., & Harrison, A. E. (1999). Do domestic firms benefit from direct foreign investment? Evidence from Venezuela. *American Economic Review*, 605–618.

Ayyagari, M., & Kosová, R. (2010). Does FDI facilitate domestic entry? Evidence from the Czech Republic. *Review of International Economics*, 18(1), 14–29.

Bai, C. E., Du, Y., Tao, Z., & Tong, S. Y. (2004). Local protectionism and regional specialization: Evidence from China's industries. *Journal of International Economics*, 63(2), 397–417.

Bates, T. (1997). Financing small business creation: The case of Chinese and Korean immigrant entrepreneurs. *Journal of Business Venturing*, 12(2), 109–124.

Batjargal, B., & Liu, M. (2004). Entrepreneurs' access to private equity in China: The role of social capital. *Organization Science*, 15(2), 159–172.

Blomström, M., & Kokko, A. (1998). Multinational corporations and spillovers. *Journal of Economic surveys*, 12(3), 247–277.

Caves, R. E. (1996). *Multinational enterprise and economic analysis*. Cambridge: Cambridge University Press.

Chang, W., & MacMillan, I. C. (1991). A review of entrepreneurial development in the People's Republic of China. *Journal of Business Venturing*, 6(6), 375–379.

Chen, C., Chang, L., & Zhang, Y. (1995). The role of foreign direct investment in China's post-1978 economic development. *World Development*, 23(4), 691–703.

Cheung, K. Y., & Lin, P. (2004). Spillover effects of FDI on innovation in China: Evidence from the provincial data. *China Economic Review*, 15(1), 25–44.

Crépon, B., & Duguet, E. (2003). Bank loans, start-up subsidies and the survival of the new firms: An econometric analysis at the entrepreneur level. Economics Working Paper Archive EconWPA, Labor and Demography (0411004).

De Backer, K., & Sleuwaegen, L. (2003). Does foreign direct investment crowd out domestic entrepreneurship? *Review of Industrial Organization*, 22(1), 67–84.

Delgado, M., Porter, M. E., & Stern, S. (2010). Clusters and entrepreneurship. *Journal of Economic Geography*, 10(4), 495–518.

Development Research Center of the State Council (DRCSC) (2004). A report on local protection in China. *Review of Economic Research*, 18, 31–38 (in Chinese).

Djankov, S., & Hoekman, B. (2000). Foreign investment and productivity growth in Czech enterprises. *The World Bank Economic Review*, 14(1), 49–64.

Djankov, S., & Murrell, P. (2002). Enterprise restructuring in transition: A quantitative survey. *Journal of Economic Literature*, 739–792.

Dollinger, M. J. (2002). *Entrepreneurship: Strategies and resources*. Upper Saddle River: Prentice Hall.

Glaeser, E. L., & Kerr, W. R. (2009). Local industrial conditions and entrepreneurship: How much of the spatial distribution can we explan? *Journal of Economics & Management Strategy*, 18(3), 623–663.

Haddad, M., & Harrison, A. (1993). Are there positive spillovers from direct foreign investment? Evidence from panel data for Morocco. *Journal of Development Economics*, 42(1), 51–74.

He, C., Wei, Y. D., & Xie, X. (2008). Globalization, institutional change, and industrial location: Economic transition and industrial concentration in China. *Regional Studies*, 42(7), 923–945.

Hexter, J. L., & Snow, J. W. (1970). An entropy measure of relative aggregate concentration. *Southern Economic Journal*, 239–243.

Javorcik, B. S. (2004). Does foreign direct investment increase the productivity of domestic firms? In search of spillovers through backward linkages. *American Economic Review*, 605–627.

Jeon, Y., Park, B. I., & Ghauri, P. N. (2013). Foreign direct investment spillover effects in China: Are they different across industries with different technological levels? *China Economic Review*, 26, 105–117.

Keuschnigg, C., & Nielsen, S. B. (2004). Taxation and venture capital backed entrepreneurship. *International Tax and Public Finance*, 11(4), 369–390.

Konings, J. (2001). The effects of foreign direct investment on domestic firms: Evidence from firm level panel data in emerging economies. *Economics of Transition*, 9(3), 619–633.

Landes, D. S. (1999). *The wealth and poverty of nations: Why some are so rich and some so poor*. New York: W.W. Norton & Company.

Lewin, A. Y., Long, C. P., & Carroll, T. N. (1999). The coevolution of new organizational forms. *Organization Science*, 10(5), 535–550.

Liao, D., & Sohmen, P. (2001). The development of modern entrepreneurship in China. *Stanford Journal of East Asian Affairs*, 1(1), 27–33.

Lin, P., Liu, Z., & Zhang, Y. (2009). Do Chinese domestic firms benefit from FDI inflow? Evidence of horizontal and vertical spillovers. *China Economic Review*, 20(4), 677–691.

Liu, X., Burridge, P., & Sinclair, P. J. (2002). Relationships between economic growth, foreign direct investment and trade: Evidence from China. *Applied Economics*, 34(11), 1433–1440.

Liu, Z. (2008). Foreign direct investment and technology spillovers: Theory and evidence. *Journal of Development Economics*, 85(1), 176–193.

Lu, J., & Tao, Z. (2010). Determinants of entrepreneurial activities in China. *Journal of Business Venturing*, 25(3), 261–273.

McMillan, J., & Naughton, B. (1992). How to reform a planned economy: Lessons from China. *Oxford Review of Economic Policy*, 130–143.

Naughton, B. (1996). *Growing out of the plan: Chinese economic reform, 1978–1993*. New York: Cambridge University Press.

North, D. C. (1990). *Institutions, institutional change and economic performance*. Cambridge: Cambridge University Press.

Schumpeter, J. A. (1934). *The theory of economic development: An inquiry into profits, capital, credit, interest, and the business cycle* (Vol. 55). Cambridge, MA: Transaction Publishers.

Tan, J., & Chow, I. H. S. (2009). Isolating cultural and national influence on value and ethics: A test of competing hypotheses. *Journal of Business Ethics*, 88(1), 197–210.

Wei, Y. D. (2001). Decentralization, marketization, and globalization: The triple processes underlying regional development in China. *Asian Geographer*, 20(1–2), 7–23.

Wei, Y. D. (2002). Beyond the Sunan model: Trajectory and underlying factors of development in Kunshan, China. *Environment and Planning A*, 34(10), 1725–1747.

Zhao, X. B., & Zhang, L. (1999). Decentralization reforms and regionalism in China: A review. *International Regional Science Review*, 22(3), 251–281.

Zhu, S., & He, C. (2013). Geographical dynamics and industrial relocation: Spatial strategies of apparel firms in Ningbo, China. *Eurasian Geography and Economics*, 54(3), 342–362.

7 The intra-metropolitan geography of entrepreneurship

A spatial, temporal, and industrial analysis (1989–2010)

Elizabeth A. Mack and Kevin Credit

Introduction

Since the Great Recession of 2008, economic development efforts are increasingly focused on fostering the generation of local businesses through entrepreneurship initiatives (Federal Reserve Atlanta, 2013). Cities across the United States are embracing downtown redevelopment plans that explicitly target entrepreneurs as a means of combating the population flight and disinvestment that continue to plague downtown areas (Federal Reserve Atlanta, 2013; Gazette-Virginia, 2014). The Downtown Project, for example, lists investment in entrepreneurs as a core component of their downtown revitalization efforts for Las Vegas (Downtown Project, 2013). In Detroit, programs such as "Thrive" and "Bizdom U" provide assistance to Detroit-based start-ups (Conlin, 2011). Unfortunately, little information about the location tendencies of start-up activity is known. In fact, most of what we know about cities and business location stems from studies evaluating the locations of existing businesses in the producer and business services sectors (Fujii and Hartshorn, 1995; Longcore and Rees, 1996; Leslie, 1997; Ó'Huallacháin and Leslie, 2007).

Prior work on new firm formation patterns across urban, rural, and suburban areas highlights the need for more spatially disaggregated analyses to understand explanatory factors behind location trends in entrepreneurial activity (Renski, 2008). Our existing understanding of the distribution of entrepreneurial activity is derived from studies at much more aggregated spatial scales, such as the metropolitan area (Lee et al., 2004), state (Parajuli and Haynes, 2012), and labor market area (Armington and Acs, 2002) levels. From an economic development and competitiveness perspective, understanding these patterns is important to unpacking a key source of competitiveness and vitality.

Comparative work across metropolitan areas is also needed to identify common trends and outliers. For example, what metropolitan areas have had exceptional success in fostering new firm formation? Conversely, what are metropolitan areas that have had little success in cultivating new firm activity? It is also important to understand where within metropolitan areas new firms are located, which will provide information about the extent to which new firms are a factor in downtown revitalization efforts. Industrial classification data can also provide anecdotal information about whether new firms are agents of industrial diversification – in terms of unrelated variety – because they represent sectors that were previously underrepresented in the metropolitan area (Frenken et al., 2007). Given the need for more work in this area, the goal of this exploratory study is to provide an initial examination of the intra-metropolitan location patterns of new businesses, as well as their temporal and industrial trends. The ten selected metropolitan study areas are geographically diverse, have different development trajectories, and distinct industrial bases. The results of this analysis highlight spatial, temporal, and industrial similarities across these diverse metropolitan areas, but also some distinct and important trends that highlight the value of a comparative approach to intra-metropolitan location studies of new firm activity.

Background

As highlighted above, very little information is available about the determinants of intra-metropolitan location patterns of entrepreneurial activity. While prior research has studied the location of entrepreneurial activity within metropolitan areas (Stough et al., 1998; Renski, 2008), the focus of these studies was the identification of spatial patterns, rather than an explanation of the factors driving these patterns.

Regional determinants of entrepreneurship

An extensive amount of research has been dedicated to documenting factors that explain variations in the distribution of entrepreneurial activity over space (Armington and Acs, 2002; Acs and Armington, 2004; Bosma and Schutjens, 2009). Demographic factors that impact regional entrepreneurial activity include age (Curran and Blackburn, 2001; Kautonen, 2008), gender (Delmar and Davidsson, 2000; Langowitz and Minniti, 2007), and educational attainment (Armington and Acs, 2002; Acs and Armington, 2004). These studies highlight that entrepreneurs tend to be young males with higher levels of educational attainment.

Economic factors that impact regional entrepreneurial activity include the unemployment rate (Storey, 1991; Bosma and Schutjens, 2009), the size of firms (Armington and Acs, 2002), and the industrial mix of existing firms (Campi et al., 2004; Acs and Armington, 2004; Delgado et al., 2010). Work evaluating the impact of unemployment has found conflicting results; some studies have found a positive association (Faber, 1999; Armington and Acs, 2002), while others have found a negative association (Storey, 1991; Audretsch and Fritsch, 1994). This is likely because unemployment may function as both a push and pull factor to pursuing entrepreneurial opportunities

(Storey, 1991). Prior work also finds that locales with smaller establishments are more conducive to entrepreneurship than those with larger establishments (Armington and Acs, 2002). Proponents of diversity suggest that it promotes information flows between people (Jacobs, 1969) and innovation (Feldman and Audretsch, 1999; Duranton and Puga, 2000). Other studies suggest that specialized economies are conducive to entrepreneurial activity (Armington and Acs, 2002; Delgado et al., 2010) because they reduce start-up costs and increase access to production inputs and complementary products (Delgado et al., 2010).

Institutional factors are important because they influence the entrepreneurial climate of places (Hwang and Powell, 2005). One example of an institutional factor that negatively impacts entrepreneurial activity is perceived administrative complexities (Grilo and Irigoyen, 2006). Other examples of factors that characterize the institutional environment include a culture that is conducive to members of the creative class (Lee et al., 2004), an environment that has a positive attitude towards entrepreneurship (Uhlaner and Thurik, 2007), and a culture that supports risk-taking (Spigel, 2012).

While the literature evaluating regional determinants of entrepreneurship is important given the locally embedded nature of entrepreneurship (Malecki, 1993; Feldman, 2001) and the importance of the regional environment in fostering entrepreneurial activity (Tödtling and Wanzenböck, 2003; Hackler and Mayer, 2008), these studies are conducted at coarser spatial scales such as states (Parajuli and Haynes, 2012) and metropolitan areas (Lee, Florida, and Acs, 2004). As mentioned previously, little information is known about the determinants of entrepreneurial activity within metropolitan areas.

Intra-metropolitan trends in business activity

While there is a wealth of literature regarding the agglomerative benefits of urban areas (Rosenthal and Strange, 2001; Nelson, 2005; Puga, 2010), there is comparatively less work on the intra-metropolitan location patterns of businesses within cities. Research related to the urban incubator hypothesis has found both evidence against (Leone and Struyk, 1976; White et al., 1993; Renski, 2008) and evidence in favor (Fagg, 1980) of this hypothesis. More recent exploratory work by Renski (2008) evaluates the location and survival of new firms across urban, suburban, and rural areas finds higher firm failure rates for new firms in central cities and favorable environments for entrepreneurial activity in suburban areas.

Studies of intra-metropolitan location patterns of existing businesses, particularly those related to producer and business services, offer some additional information about the factors behind the location patterns uncovered. These pieces find evidence of strategic suburbanization of firms over time (Fujii and Hartshorn, 1995; Longcore and Rees, 1996; Leslie, 1997). They also highlight that these patterns represent a tension between cost and information needs. Leslie (1997) highlights that the ability to use information technology (IT) has enabled firms to decentralize for cost purposes. Ó'Huallacháin and Leslie (2007) suggest, however, that firms may choose to remain centrally located because of information externalities. Other reasons cited for centralized location tendencies include the need for a highly educated labor force, access to inputs, and the ability to sell outputs (Coffey and Shearmur, 1997).

While valuable, prior work examining the regional determinants of entrepreneurial activity has analyzed this topic at either too coarse a spatial scale to understand what is happening within metropolitan areas, or has focused on *existing* business patterns within cities. The work that has analyzed new firm survival within metropolitan areas has done so for a limited time period for just two sectors, manufacturing and advanced business and professional services (Renski, 2008). More work is certainly needed to understand trends in multiple sectors over longer periods of time in a comparative context. The present study analyzes 20 different industry sectors over 21 years for ten U.S. metropolitan areas.

Geographic and temporal context

To provide more fine-grained information about the spatial, temporal, and industrial patterns of new businesses within metropolitan areas, this study has assembled a unique database of new establishment information for ten metropolitan areas around the United States. Figure 7.1 highlights the distribution of these places around the country. These metropolitan areas were selected because they represent a geographically and industrially diverse range of places. They also have varied levels of entrepreneurial activity. Austin, Texas; Boulder, Colorado; Raleigh, North Carolina; San Jose, California (Silicon Valley); and Boston, Massachusetts are all historically active centers of entrepreneurial activity (Saxenian, 1994; Feld, 2012). Fayetteville, Arkansas and Omaha, Nebraska were selected because they are located in comparatively rural, Midwestern settings and are not regarded as hubs of entrepreneurial activity.

Phoenix and Kansas City also represent locales for which we know comparatively little about entrepreneurship, but may be transitioning slowly to becoming important hubs of entrepreneurial activity. Since its founding at the turn of the twentieth century, Phoenix has expanded rapidly and is now one of the largest metropolitan areas in the country (Shermer, 2013). In recent years, the Phoenix economy has tracked national business cycles closely and suffered economically from the 2008 recession and ensuing foreclosure crisis (Reagor et al., 2011; Reagor, 2013). Like Phoenix, Kansas City is also a non-traditional hub of entrepreneurial activity; it is a Midwestern city that developed around the railroad and meatpacking industries (KCSHS, 2014; Onasch, 2014). Since its meatpacking years, the metropolitan area has developed related specializations in life sciences, telecommunications, and information technology (Mayer, 2013). Today, Kansas City is regarded as a second-tier entrepreneurial region that is founded on these industrial ties (Mayer, 2013).

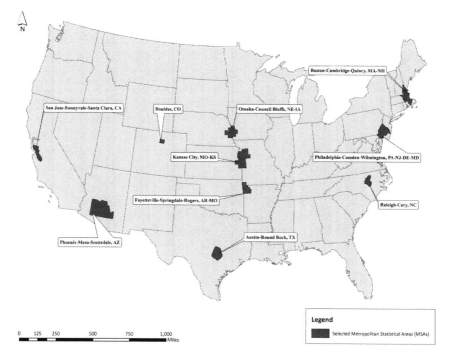

Figure 7.1 Ten selected metropolitan areas.

The tenth metropolitan area, Philadelphia, was selected because it is one of the more historic urban centers in the United States, which also has an unfortunate legacy of crime, disinvestment, and unemployment (Field, 2014).

This chapter analyzes these ten metropolitan areas over a 21-year period (1989–2010), which was selected because it represents a particularly interesting time for entrepreneurial activity in the United States. This time period not only includes the "dot-com" boom and bust (Ayres and Williams, 2004), but also the years associated with the housing bubble that led to the 2008 stock market crash and associated Great Recession of 2008 (Fairlie, 2013). Aside from important business cycles, this was also a time of tremendous advancements in information and communication technologies (ICTs), including the Internet, social media, and smart phones. In addition to the business environment and business opportunities associated with innovations in ICTs, which have likely favored small businesses (Wennekers et al., 2005), these innovations also offer small businesses more virtual visibility than ever before. The rise of services in a post-industrial economy has also provided more opportunities for business ownership because of the powerful potential of exploiting new market niches that have opened up with leaner, more customized product and service offerings (Wirtz et al., 2015; Wennekers et al., 2005).

New firm data

In this analysis, entrepreneurial activity is defined as new establishments that have not relocated from within the region or from out of state, as identified in the National Time Series Establishment (NETS) database (Walls & Associates, 2014). This is a proprietary, point-level database that contains detailed information about all establishments in the ten metropolitan areas of interest for the years 1989–2010. It also contains details about firm industry membership for both the Standard Industrial Classification (SIC) system and the North American Industrial Classification System (NAICS), as well as the start date and end date of the establishment.[1] The formation of new establishments is indicated by the start date as reported by the establishment.[2] The number of establishments used in this chapter is based on the precedent set by prior studies which highlights that other measures such as entry rates (the number of new businesses per population or people in the labor force) may not capture important local information that influence new firm formation (Renski, 2008; Acs and Audretsch, 2010).

Before proceeding to a discussion of the results, there are some important points to be made about the use of establishment data. First, establishments *are not the same entities as companies*. In fact, the U.S. Census Bureau defines establishments as the physical location at which business activity takes place (U.S. Census Bureau, 2014). Second, entrepreneurial activities consist of a range of actions taken to start a business and are a multidimensional concept that is difficult to encapsulate in a single measure (Renski, 2008). These activities may include conversations with other aspiring entrepreneurs, meetings with venture capitalists, and starting a business from home.

Thus, in the context of this study, the data speak to entrepreneurs that have started a physical place of business. It does not consider aspiring entrepreneurs at other stages in the start-up process that are not included in data reported by public and private statistical agencies. Defining entrepreneurial activity as new establishments also does not necessarily isolate the kinds of new businesses most associate with start-ups (i.e., high-technology, bio-technology, or Internet companies). Instead, the data used in this study provide a more inclusive definition of entrepreneurship that also includes traditional and perhaps necessity-based entrepreneurial activity. The implications of this definition for the analytical results will be discussed at length toward the close of this chapter, as will directions for future research.

Temporal trends in start-ups

Figure 7.2 presents times series trends in new establishment activity per 100,000 people in the ten metropolitan areas of interest. This time series highlights steady growth in new establishment starts over time, and also indicates that new establishment starts are associated with both boom and recession periods. The dot-com stock market bubble peaked in 2000 and

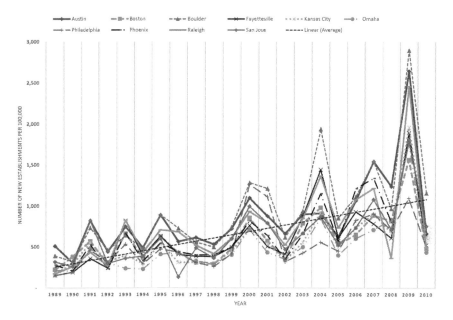

Figure 7.2 New establishments per 100,000 population

began to collapse in 2001 (Google Finance, 2015); this corresponds with the relative peak in new establishments per capita in all of the selected regions in 2000, followed by a subsequent drop. Interestingly, however, the data for the Great Recession, which began at the very end of 2007 and extended through the middle of 2009 (NBER, 2012), show a different trend. While all metros demonstrate a relative drop in per capita start-up rates in 2008, the numbers of new establishments peak (for the entire time series) in 2009, perhaps reflecting the contrasting push and pull factors that business cycle downturns create for new business starts.

Figure 7.2 also demonstrates the fact that well-known hubs of entrepreneurial activity such as Austin, Boulder, and Raleigh confirm their status as some of the places with the highest levels of new establishment activity; although it should be noted that some of the high per capita levels are almost certainly due to smaller populations in these regions. Even so, Austin ranks first in 1989 for per capita establishment starts, and then consistently in the top three regions in the years thereafter. A similar situation is true for Boulder, which reliably ranks in the top two of the selected regions. Interestingly, Raleigh ranked first in 1993, and maintains a ranking close to the top over the 1989–2010 study period. Not only is the city of Raleigh the state capital of North Carolina, but the metropolitan area is located just 18 to 21 miles from the Research Triangle Park (Kroll, 2015) which is a key hub of research, innovation, and entrepreneurial activity in the United

States (Feldman and Lowe, 2015). This status and the metro area's growing reputation for a good quality of life (Kroll, 2015) are likely explanations for the overall growth of Raleigh establishment starts over the last two decades.

Locales that ranked among the lowest in new establishment activity per capita include Philadelphia, Pennsylvania; Phoenix, Arizona; and Fayetteville, Arkansas. Again there is year-to-year variability in these rankings, as with the top metropolitan areas for new establishment activity. Of these metropolitan areas, Philadelphia and Omaha trade places for the locales with the fewest amount of new establishments per 100,000 people in this 21-year period. Although Philadelphia has been making efforts to foster new business activity by making investments in small business development (SGA, 2011), these efforts are a challenging prospect for the city given its decades of disinvestment and urban problems. Philadelphia is not only considered the poorest big city in the United States, but it also has a poverty rate (12.2 percent) that is double the U.S. average of 6.3 percent (Lubrano, 2014). This poverty issue, which is also a workforce problem that provides a disincentive for businesses to locate in the city (Lubrano, 2014), combined with the recent uptick in gun-related and violent crimes (Odom, 2015), makes growing and attracting businesses an ongoing challenge for the metropolitan area.

The results for Omaha are not necessarily surprising either. This Nebraska metropolitan area is located in a comparatively remote portion of the United States compared to places like Austin and San Jose. Omaha may be a metropolitan area to watch in coming years, however, given the downtown redevelopment efforts underway and its strides in cultivating a local music scene. In particular, the 56,000 square foot Slowdown urban redevelopment project initiated in 2007 in the blighted North Downtown area of Omaha has helped bring multiple events and concerts to the city and infused the city with a sense of vitality (Seman, 2010).

Of the places with lower levels of new establishment activity, Fayetteville and Phoenix are some of the more interesting regions. Both metropolitan areas have cyclical trends to their new establishment activity. In 1990 Phoenix ranked seventh among the ten metropolitan areas of interest and tenth in 1992. But since this year, the metropolitan area has made some upward strides in new establishment activity, peaking at number one in 2006, with five top-five finishes since 2004. In 2011, the Kauffman Foundation ranked Phoenix first in the country for entrepreneurial activity (Hoover, 2012).[3] According to Kauffman's data about start-ups, Phoenix had 520 start-ups per 100,000 adults compared to a national average of 320 per 100,000 (Hoover, 2012). There are several potential explanations for this trend. One, Phoenix, Arizona is home to the largest research university in the United States, Arizona State (NCES, 2012). Two, the metropolitan area is in close proximity to California and Silicon Valley. Three, the state has taken a growth-machine approach to developing its entrepreneurial ecosystem with some success; although, the metropolitan area still lacks an organic flavor to the ecosystem (Mack and

Mayer, 2015) generated by a regional culture of entrepreneurship (Saxenian, 1994; Feld, 2012) and visible, successful start-ups (Isenberg, 2011).

The temporal trend in new establishment activity for Fayetteville is somewhat comparable to Phoenix, although more recently, activity has trended downward. In 1990 and 1991 Fayetteville ranked last in new establishment starts per capita, but second in 1993 and 2004. Since this 2004 peak, Fayetteville ranked ninth in 2007 and 2008 and fifth in 2009 and 2010. The cyclical nature of new establishment starts here may be tied to the fortune of Walmart, whose headquarters are located in Bentonville, Arkansas, about 35 miles to the north of Fayetteville. Between 1988 and 2000, Walmart's CEO David Glass grew company profits to make it one of the world's largest retailers through innovations in supply chains and logistics (Friedman, 2005). In fact, the area around Walmart headquarters is known as "Vendorville" because vendors around the globe have set up shop to be close to headquarters (Friedman, 2005). Establishments have likely sprung up to service these vendors who demand products and services while they are in town. New establishment starts in the region are also likely tied to Tyson Foods, which has several food plants located in southeastern Arkansas (Tyson Foods, 2014).

Industrial trends in start-ups

Although the differences in temporal trends between metropolitan areas are not necessarily surprising, the industry similarities in new establishments across metropolitan areas is unexpected. For nine out of the ten metropolitan areas, five two-digit North American Based Industrial Classification System[4] (NAICS) industries constituted the top five industries over the entire 21-year study period. These five industries are: Construction (NAICS 23), Retail Trade (NAICS 44–45), Professional, Scientific, and Technical Services (NAICS 54), Administrative and Support and Waste Management and Remediation Services (ASWMRS) (NAICS 56), and Other Services (except Public Administration) (NAICS 81). The only metropolitan area that deviates from this trend (and even then, only slightly) is Omaha. In this metropolitan area, health care and social assistance is also a top industry, in addition to the five industries the other metropolitan areas have in common. The large number of new business starts in this industry are perhaps a reflection of several large healthcare companies in the metropolitan area, which include: the Methodist Health Center, the Nebraska Medical Center, the University of Nebraska Medical Center, and Children's Hospitals & Medical Center (GOEDP, 2014).

Of the five industries that all metropolitan areas have in common, perhaps the more interesting are the professional/scientific/technical services (PSTS) and administrative/support/waste management and remediation services (ASWMRS). PSTS is interesting because it contains establishments dealing in creative economy, knowledge-oriented production processes such as computer systems design and related services (NAICS 5415), management, scientific,

(a)

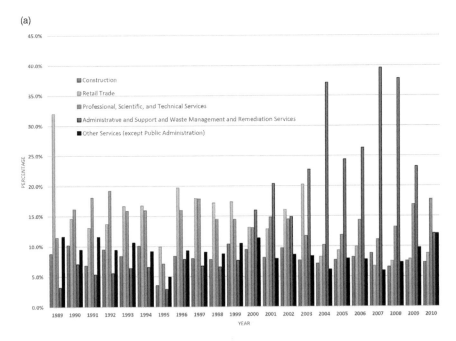

Figure 7.3a Austin new establishments by industry

and technical consulting services (NAICS 5416), and scientific research and development services (NAICS 5417). Establishment starts in administrative and waste (ASWRS), which range from employment services to waste treatment and disposal, are interesting because of the tremendous growth in this industry category. Figures 7.3a and 7.3b contain two sets of bar graphs, demonstrating this trend in two very different metropolitan areas: Austin and Omaha. Although Austin is widely regarded as a tech hub (Feld, 2012) and Omaha an insurance and financial service hub, both exhibit similar trends in ASWRS new business starts. In Austin, this industry accounted for just 3.2 percent of all new establishments in 1989 and almost 40 percent of new establishments in both 2007 and 2008. In Omaha a similar trend is apparent. In 1989, ASWRS accounted for 7.7 percent of all new establishments and almost 43 percent and 31 percent of new establishments in 2007 and 2008, respectively. This trend is common across all ten metropolitan areas and clearly highlights the post-industrial services orientation of new businesses in the later portions of the study period.

Within the professional/scientific/technical services at the four-digit NAICS level, there are distinct trends in the types of businesses that are being created. For all metropolitan areas expect Boston, Omaha, and Phoenix, growth in this sector was driven by other professional, scientific, and technical services, which contains businesses involved in market research and opinion polling,

(b)

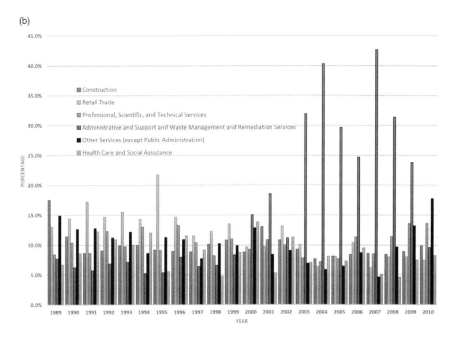

Figure 7.3b Omaha new establishments by industry

photographic services, translation and interpretation services, and veterinary services. Growth in this sector was also driven by establishments providing management, scientific, and technical services in all of the metropolitan areas save Omaha and Philadelphia. The explosive growth in the administrative/support/waste management and remediation services in a majority of metropolitan areas was driven by growth in Other Support Services, which includes packaging and labeling services, as well as convention and trade show organization.

A related question to the industry trends discussed above is the relative diversity of new establishment starts over time: are they more or less evenly distributed across industries, or are they highly concentrated in specific industries? Figure 7.4 display trends in the Herfindahl index computed for new establishments between 1999 and 2010. Low values of the index correspond to diversity in new business starts while high values indicate specialization or the clustering of business starts in particular industries. The time period presented in Figure 7.4 was selected because it represents a point in time when start-ups began to transition from relative diversity to clustering in select industries. Prior to 2003, start-ups were more or less evenly distributed across industries. Starting in about 2003, however, clustering began to take place in specific industries, most notably in the administrative/support/ waste management and remediation services. This rise in starts for ASWRS

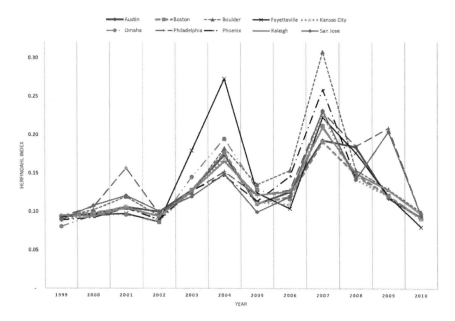

Figure 7.4 Herfindahl Index

largely accounts for the spikes in specialization in 2001, 2004, and 2007. In Fayetteville, for example, this industry category accounted for about half (44 percent) of all new businesses in 2007. Even in tech-centered Boulder, ASWRS accounted for 53.2 percent of business starts in that year, which represents a change in focus from professional, scientific, and technical services in previous years.

Spatial trends in start-ups

Aside from the temporal and industrial trends associated with start-ups in each of the ten metropolitan areas, another key question to consider is the spatial distribution of new establishments over time. This would help answer the question of whether there appears to be path dependence in the location of start-ups. Or, in other words, does new establishment activity appear to be concentrated in particular portions of these metropolitan areas, or does it appear to be dispersed across metropolitan areas? An examination of kernel density estimate maps[5] created for each metropolitan area at ten-year intervals (1989, 1999, and 2009[6]) highlights a diffusion of new establishment activity from one central or multiple cores. In Phoenix, Arizona, for example, there is a noticeable exurban pattern to new establishment growth. This trend is most prominent in exurban cities of the metropolitan area including Surprise, Avondale, Cave Creek, Fountain Hills, and Queen Creek. Figure 7.5 shows

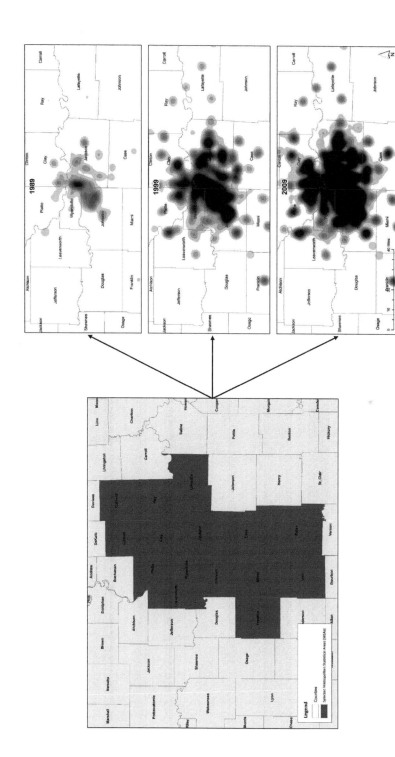

Figure 7.5 Spatio-temporal trends in new establishment activity: Kansas City

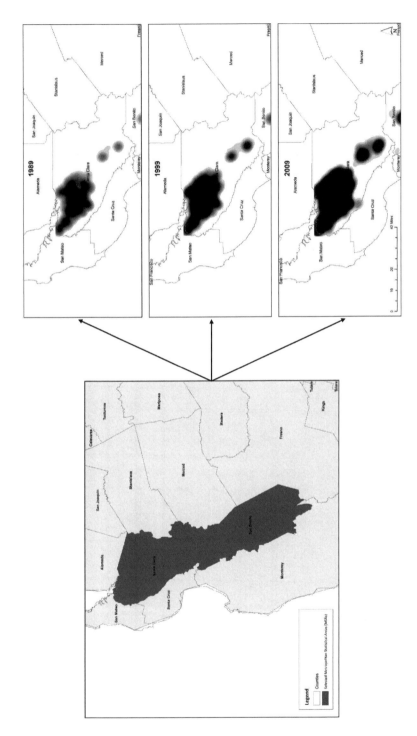

Figure 7.6 Spatio-temporal trends in new establishment activity: San Jose

Kansas City and highlights that, since 1989, activity in the central city core, Johnson County and Jackson County has intensified and expanded outward to neighboring counties. By 2009, the expansions grew to include Leavenworth, Platte, Clay, Cass, and even Miami Counties. Similarly, core hotspots of new establishment activity in Boulder, Colorado have diffused south and north toward Denver and Fort Collins, respectively. The one notable exception to this general pattern of outward expansion is San Jose. Interestingly, in this metropolitan area, new business starts merely intensify around the existing central core of new business activity. Figure 7.6 shows the apparent spatial inertia in the distribution of new establishments in the most established tech hub in the United States, San Jose.

Discussion and conclusion

The goal of this study was to analyze the intra-metropolitan location patterns of entrepreneurial activity in a comparative context using several metropolitan areas with diverse urban forms, industrial histories, and levels of entrepreneurial activity. The results of the analysis highlight clear temporal, spatial, and industrial trends in new establishment starts across metropolitan areas.

Other researchers have noted that we are in the age of entrepreneur-ship and small business (Drucker, 1984; Audretsch and Thurik, 2001), and this is reflected in the temporal trends of new establishment starts over time. Since 1989, there is a clear upward trend in the number of new establishments in metropolitan areas. This is also reflected in the increasing intensification of new establishment activity within existing centers of establishment activity.

Business activity clearly follows an outward pattern of metropolitan expansion. This is particularly interesting from an urban perspective, because in the 1990s, suburbanization was already in full swing (Preston and McLafferty, 1999; Lopez and Hynes, 2003). The continued outward expansion demonstrated by the location patterns of new establishments since 1989 clearly shows the emergence of a second tier of suburban expansion toward the exurbs (Nelson and Dueker, 1990; Theobald, 2001) over the time period of this study.

In addition to clear spatial and temporal trends in new establishment starts, there are also strong industrial trends. For this 21-year period, five industries accounted for the bulk of new establishment starts. These industries were primarily in the services and retail sectors. Of the service starts, many of these were in packaging and labeling services and trade show and convention organization. This could reflect the growth of ICT-oriented businesses and Internet retailers that rely on shipping and "just-in-time" logistical services such as Amazon, as well as the role of temporary clusters such as trade fairs and conventions in providing venues for intense knowledge exchange in a global economy (Maskell et al., 2006). However, the direct cause is hard to tell directly from the present analysis, and this interesting trend deserves further investigation.

That said, the results of this analysis are expected to vary regionally and according to different interpretations of entrepreneurship. This study uses a definition of entrepreneurship derived from Acs and Audretsch (2010), who conceptualized entrepreneurship as all new business activity, irrespective of industry membership or size. Studies that focus on businesses in particular industries or businesses with high-growth prospects are likely to find deviations in the trends uncovered in the present analysis. Irrespective of this anticipated sources of variation in results, we hope that the present study will encourage future research about intra-metropolitan patterns of new firm activity in the United States and other cities around the globe.

Acknowledgement

The research in chapter 7 was funded by the Ewing Marion Kauffman Foundation grant #20130782.

Notes

1 For more details about the NETS database, readers are referred to Walls & Associates (2014).
2 Where start data were not available, the first year the establishment appeared in the database was used as a proxy for the start date.
3 2011 data were not available for Phoenix and thus the analysis for Phoenix in this study stops at 2009 which means no comparable data are available to compare directly to the data from the Kauffman Foundation.
4 The North American Industrial Classification Systems or NAICS for short, is a common means of classifying industry activity for all three countries in North America (Canada, the United States, and Mexico). NAICS groups industries by their similarity in production activity rather than product similarity as did its antecedent, the Standard Industrial Classification (SIC) system (Walker and Murphy, 2001).
5 These kernel density maps are identical in terms of the search radius, kernel, raster grid size, map scale, map extent, and symbology levels for displaying kernel intensity.
6 Due to data availability limits, the maps generated for Phoenix actually pertain to the years 1990, 2000, and 2009.

References

Acs, Z. J. and C. Armington (2004). The impact of geographic differences in human capital on service firm formation rates. *Journal of Urban Economics*, 56(2): 244–278.
Acs, Z. J. and D. B. Audretsch (2010). *Introduction. In Handbook of entrepreneurship research, 2nd edition.* New York: Springer, 1–19.
Armington, C. and Z. J. Acs (2002). The determinants of regional variation in new firm formation. *Regional Studies*, 36(1): 33–45.
Audretsch, D. B. and M. Fritsch (1994). The geography of firm births in Germany. *Regional Studies*, 28(4): 359–365.

Audretsch, D. B. and A. R. Thurik (2001). *Capitalism and democracy in the 21st century: from the managed to the entrepreneurial economy.* Journal of Evolutionary Economics, 10: 17–34.

Ayres, R. U. and E. Williams (2004). The digital economy: where do we stand? *Technological Forecasting and Social Change*, 71(4): 315–339.

Bosma, R. Niels and V. Schutjens (2009). Mapping entrepreneurial activity and entrepreneurial attitudes in European regions. *International Journal of Entrepreneurship and Small Business*, 7(2): 191–213.

Campi, M., T. Costa, A. S. Blasco, and E. V. Marsal (2004). The location of new firms and the life cycle of industries. *Small Business Economics*, 22(3–4): 265–281.

Coffey, W. J. and R. G. Shearmur (1997). The growth and location of high order services in the Canadian urban system, 1971–1991. *The Professional Geographer*, 49(4): 404–418.

Conlin, J. (2011). Detroit pushes back with young muscles. Retrieved from: www.nytimes.com/2011/07/03/fashion/the-young-and-entrepreneurial-move-to -downtown-detroit-pushing-its-economic-recovery.html?pagewanted=all&_r=0 (accessed July 2011).

Curran, J. and R. A. Blackburn (2001). Older people and the enterprise society: age and self-employment propensities. *Work, Employment & Society*, 15(4): 889–902.

Delgado, M., M. E. Porter, and S. Stern (2010). Clusters and entrepreneurship. *Journal of Economic Geography*, 10(4): 495–518.

Delmar, F. and P. Davidsson (2000). Where do they come from? Prevalence and characteristics of nascent entrepreneurs. *Entrepreneurship & Regional Development*, 12(1): 1–23.

Downtown Project (2013). Downtown Las Vegas. Retrieved from: http://downtownproject.com/ (accessed July 2015).

Drucker, P. F. (1984). Our entrepreneurial economy. *Harvard Business Review*, 62(1): 58–64.

Duranton, G. and D. Puga (2000). Diversity and specialisation in cities: why, where and when does it matter? *Urban Studies*, 37(3): 533–555.

Faber, H. S. (1999). Alternative and part-time employment arrangements as a response to job loss. *Journal of Labor Economics*, 17(4): 142–169.

Fagg, J. J. (1980). A re-examination of the incubator hypothesis: A case study of Greater Leicester. *Urban Studies*, 17(1): 35–44.

Fairlie, R. W. (2013). Entrepreneurship, economic conditions, and the Great Recession. *Journal of Economics and Management Strategy*, 22(2): 2017–231.

FederalReserveAtlanta(2013).Revitalizingdowntownthroughentrepreneurship.Retrieved from: www.frbatlanta.org/podcasts/transcripts/economicdevelopment/130322_leach_marinucci.cfm?d=1&series=Podcasts (accessed: March 2013).

Feld, B. (2012). *Startup communities: building an entrepreneurial ecosystem in your city*. Hoboken: John Wiley & Sons, Inc.

Feldman, M. P. (2001). The entrepreneurial event revisited: firm formation in a regional context. *Industrial and Corporate Change*, 10(4): 861–891.

Feldman, M. P. and D. B. Audretsch (1999). Innovation in cities: science-based diversity, specialization and localized competition. *European Economic Review*, 43(2): 409–429.

Feldman, M. and N. Lowe (2015). Triangulating regional economies: realizing the promise of digital data. *Research Policy*. DOI: http://dx.doi.org/10.1016/j.respol.2015.01.015.

Field, A. (2014). Philly's "super-group" startup accelerator targets urban problems. Retrieved from: www.forbes.com/sites/annefield/2014/01/05/phillys-super-group-startup-accelerator-targets-urban-problems/ (accessed January 2014).

Frenken, K., F. Van Oort, and T. Verburg (2007). Related variety, unrelated variety and regional economic growth. *Regional Studies*, 41(5): 685–697.

Friedman, T. L. (2005). *The world is flat: a brief history of the twenty-first century.* New York: Farrar, Straus and Giroux.

Fujii, T. and T. A. Hartshorn (1995). The changing metropolitan structure of Atlanta, Georgia: locations of functions and regional structure in a multinucleated urban area. *Urban Geography*, 16(8): 680–707.

Gazette-Virginian (2014). Destination downtown, town of South Boston receive awards for revitalization efforts. Retrieved from: http://yourgv.com/index.php/news/government/10067-destination-downtown-town-of-south-boston-receive-awards-for-revitalization-efforts (accessed March 2014).

GOEDP (2014). Greater Omaha major employers. *Greater Omaha Economic Development Partnership*. Retrieved from: www.selectgreateromaha.com/Omaha/media/Omaha/Why%20Here/Major%20Employers/GreaterOmahaTop100LargestEmployerList.pdf?ext=.pdf (accessed April 2014).

Google Finance (2015). NASDAQ Composite – March 1, 2000 to March 30, 2000. Retrieved from: www.google.com/finance/historical?cid=13756934&startdate=march+01%2C+2000&enddate=march+30%2C+2000&num=30&ei=xRbbUpC2CJPzqQH_GA (accessed September 2015).

Grilo, I. and J. Irigoyen (2006). Entrepreneurship in the EU: to wish and not to be. *Small Business Economics*, 26(4): 305–318.

Hackler, D. and H. Mayer (2008). Diversity, entrepreneurship, and the urban environment. *Journal of Urban Affairs*, 30(3): 273–307.

Hoover, K. (2012). Arizona No. 1 for startups, Kauffman study finds. *Phoenix Business Journal*, March 19, 2012. Retrieved from: www.bizjournals.com/phoenix/blog/business/2012/03/arizona-no-1-for-startups-kauffman.html (accessed: September 2015).

Hwang, H. and W. W. Powell (2005). Institutions and entrepreneurship. In *Handbook of entrepreneurship research*. Boston: Springer US, 201–232.

Isenberg, D. (2011). The entrepreneurship ecosystem strategy as a new paradigm for economic policy: principles for cultivating entrepreneurship. *Babson Entrepreneurship Ecosystem Project*. Retrieved from: www.wheda.com/uploadedFiles/Website/About_Wheda/Babson%20Entrepreneurship%20Ecosystem%20Project.pdf (accessed September 2015).

Jacobs, J. (1969). *The economy of cities*. New York: Random House.

Kautonen, T. (2008). Understanding the older entrepreneur: comparing third age and prime age entrepreneurs in Finland. *International Journal of Business Science and Applied Management*, 3(3): 3–13.

KCSHS (2014). The Kansas City Southern Lines. *The Kansas City Historical Society*. Retrieved from: www.kcshs.org/schedule/subs/images/history/kcs_hist.htm (accessed August 2015).

Kroll, D. (2015). 7 reasons it's finally time to live in Research Triangle Park. *Forbes*. Retrieved from: www.forbes.com/sites/davidkroll/2014/02/04/7reasonsitsfinallytimetoliveinresearchtrianglepark/ (accessed February 4, 2015).

Langowitz, N. and M. Minniti (2007). The entrepreneurial propensity of women. *Entrepreneurship Theory and Practice*, 31(3): 341–364.

Lee, S. Y., R. Florida, and Z. Acs (2004). Creativity and entrepreneurship: a regional analysis of new firm formation. *Regional Studies*, 38(8): 879–891.

Leone, R. A. and R. Struyk (1976). The incubator hypothesis: evidence from five SMSAs. *Urban Studies*, 13(3): 325–331.

Leslie, D. (1997). Abandoning Madison Avenue: the relocation of advertising services in New York City. *Urban Geography*, 18(7): 568–590.

Longcore, T. R. and P. W. Rees (1996). Information technology and downtown restructuring: the case of New York City's financial district. *Urban Geography*, 17(4): 354–372.

Lopez, R. and Hynes, H. P. (2003). Sprawl in the 1990s: measurement, distribution, and trends. *Urban Affairs Review*, 38(3): 325–355.

Lubrano, A. (2014). Philadelphia rates highest among top 10 cities for deep poverty. Philly.Com. Retrieved from: http://articles.philly.com/20140926/news/54322611_1_ deeppovertypovertylinesouthphiladelphia (accessed September 24, 2015).

Mack, E. and H. Mayer (2015). The evolutionary dynamics of entrepreneurial ecosystems. *Urban Studies*. DOI: 10.1177/0042098015586547.

Malecki, E. J. (1993). Entrepreneurship in regional and local development. *International Regional Science Review*, 16(1–2): 119–153.

Maskell, P., H. Bathelt, and A. Malmberg (2006). Building global knowledge pipelines: the role of temporary clusters. *European Planning Studies*, 14(8): 997–1013.

Mayer, H. (2013). *Entrepreneurship and innovation in second tier regions*. Cheltenham, UK and Northampton, MA: Edward Elgar Publishing.

National Bureau of Economic Research (NBER) (2012). US business cycle expansions and contractions. Retrieved from: www.nber.org/cycles/US_Business_Cycle_ Expansions_and_Contractions_20120423.pdf (accessed September 2015).

National Center for Education Statistics (NCES) (2012). Table 312.10. Enrollment of the 120 largest degree-granting college and university campuses, by selected characteristics and institution: Fall 2012. Retrieved from: https://nces.ed.gov/ programs/digest/d13/tables/dt13_312.10.asp (accessed September 2015).

Nelson, M. K. (2005). Rethinking agglomeration economies and the role of the central city: the public accounting industry in Chicago and Minneapolis-St. Paul. *Journal of Planning Education and Research*, 24(3): 331–341.

Nelson, A. C. and K. J. Dueker (1990). The exurbanization of America and its planning policy implications. *Journal of Planning Education and Research*, 9(2): 91–100.

Odom, V. (2015). Philadelphia violent crime rate on the rise. 6Abc.Com News. Retrieved from: http://6abc.com/news/philadelphiaviolentcrimerateontherise/927865/ (accessed August 13, 2015).

Ó'Hallacháin, B. and T. F. Leslie (2007). Producer services in the urban core and suburbs of Phoenix, Arizona. *Urban Studies*, 44(8): 1581–1601.

Onasch, B. (2014). Packinghouse unionism in Kansas City. Retrieved from: www. kclabor.org/kc150june.htm (accessed July 2015).

Parajuli, J. and K. E. Haynes (2012). Broadband Internet and new firm formation: a U.S. perspective. *George Mason School of Public Policy* Research Paper No. 2013-03.

Preston, V. and S. McLafferty (1999). Spatial mismatch research in the 1990s: progress and potential. *Papers in Regional Science*, 78(4): 387–402.

Puga, D. (2010). The magnitude and causes of agglomeration economies. *Journal of Regional Science*, 50(1): 203–219.

Reagor (2013). Phoenix area sees dramatic drop in empty houses. Retrieved from: www.usatoday.com/story/money/personalfinance/2013/09/22/phoenix-area-sees-dramatic-drop-in-empty-houses/2848001/ (accessed July 2015).

Reagor, Dempsey, and Kongi (2011). Phoenix-area real estate collapse echoed troubles. Retrieved from: www.azcentral.com/business/realestate/articles/2011/10/09/20111009phoenix-real-estate-foreclosure-crisis-economy.html (accessed July 2015).

Renski, H. (2008). New firm entry, survival, and growth in the United States: a comparison of urban, suburban, and rural areas. *Journal of the American Planning Association*, 75(1): 60–77.

Rosenthal, S. S. and W. C. Strange (2001). The determinants of agglomeration. *Journal of Urban Economics*, 50(2): 191–229.

Saxenian, A. (1994). *Regional advantage: culture and competition in Silicon Valley and Route 128*. Cambridge, MA: Harvard University Press.

Seman, M. (2010). How a music scene functioned as a tool for urban redevelopment: a case study of Omaha's Slowdown project. *City, Culture, and Society*, 1(4): 207–215.

SGA (2011). Philadelphia: more than just good cheese steaks. *Smart Growth America*. Retrieved from: www.smartgrowthamerica.org/2011/10/28/philadelphia-more-than-just-good-cheese-steaks (accessed October 2011).

Shermer, E.T. (2013). *Sunbelt capitalism: Phoenix and the transformation of American politics*. Philadelphia: University of Pennsylvania Press.

Spigel, Ben. (2012). The sources of regional variation in Canadian self-employment. *International Journal of Entrepreneurship and Small Business*, 15(3): 340–361.

Storey, D. J. (1991). The birth of new firms – does unemployment matter? A review of the evidence. *Small Business Economics*, 3(3): 167–178.

Stough, R. R., K. E. Haynes, and H. S. Campbell, Jr. (1998). Small business entrepreneurship in the high technology services sector: an assessment for the edge cities of the US national capital region. *Small Business Economics*, 10(1): 61–74.

Theobald, D. M. (2001). Land-use dynamics beyond the American urban fringe. *Geographical Review*, 91(3): 544–564.

Tödtling, F. and H. Wanzenböck (2003). Regional differences in structural characteristics of start-ups. *Entrepreneurship & Regional Development*, 15(4): 351–370.

Tyson Foods (2014). Locations: Tyson Foods across the U.S. Retrieved from: www.tysonfoods.com/Our-Story/Locations.aspx (accessed July 2015).

Uhlaner, L. and R. Thurik (2007). Postmaterialism influencing total entrepreneurial activity across nations. *Journal of Evolutionary Economics*, 17(2): 161–185.

U.S. Census Bureau (2014). Definitions. Retrieved from: www.census.gov/econ/susb/definitions.html (accessed August 2015).

Walls & Associates (2014). National Establishment Time-Series (NETS) Database: Database Description. Retrieved from: https://msbfile03.usc.edu/digitalmeasures/cswenson/intellcont/NETS%20Database%20Description2008-1.pdf (accessed August 2015).

Walker, J. A. and J. B. Murphy (2001). Implementing the North American Industry Classification System at BLS. *Monthly Lab. Rev.*, 124: 15.

Wennekers, S., A. Van Wennekers, R. Thurik, and P. Reynolds (2005). Nascent entrepreneurship and the level of economic development. *Small Business Economics*, 24(3): 293–309.

White, S. B., L. S. Binkley, and J. D. Osterman (1993). The sources of suburban employment growth. *Journal of the American Planning Association*, 59(2): 193–204.

Wirtz, Jochen, S. Tuzovic, and M. Ehret (2015). Global business services: increasing specialization and integration of the world economy as drivers of economic growth. *Journal of Service Management*, 26(4): 565–587.

8 Are start-ups the same everywhere?

The urban–rural skill gap in Swedish entrepreneurship

Martin Andersson, Sierdjan Koster and Niclas Lavesson

Introduction

There is a solid consensus that entrepreneurship, measured as the start-up of a new firm, is an important driver of regional economic development and employment generation in particular. At the same time, however, the exact mechanisms governing job creation from entrepreneurship are not fully clear and empirical evidence on the matter is not univocal (Fritsch, 2013). As a case in point Fritsch and Schroeter (2011) and Fritsch and Noseleit (2013) show that there are significant differences across regions in terms of the impact of new firm formation on regional employment growth. The strength of the effect broadly follows the urban hierarchy. The positive effect of start-ups on employment are larger in cities than they are elsewhere (see, for example, Van Stel and Suddle, 2008).

One possible explanation for differences in the effects of new firm formation across space are spatial differences in the characteristics of start-ups. Firm level studies have established that only a very specific subset of all new firms are able to provide a sustained number of new jobs (Acs, 2011). If spatial sorting processes are present for types of start-ups with above-average performance, the composition of the local start-up population can be an important mechanism governing spatial variations in the impact of entrepreneurship on regional employment. Although there is now a large literature addressing the spatial patterns of start-ups, the heterogeneity in the types of start-ups has been largely ignored. That is, studies mainly focus on the *frequency* of start-ups but tend to neglect the characteristics of start-ups.

This chapter contributes to the literature by presenting an explorative analysis of spatial variations in the characteristics of start-ups in Sweden. We characterize start-ups in terms of the labour market backgrounds of the entrepreneurs involved (*ex-ante* characteristics) and information on the firm after start-up (*ex-post*) including size, educational level of the employees and industry. By focusing on the characteristics of start-ups we aim to contribute to a more nuanced discussion of entrepreneurship that builds on the premise that there is a need to acknowledge the fact that start-ups are truly heterogeneous (Henrekson and Sanandaji, 2014). Specifically, we explore the idea that the increased employment effect of start-ups along the urban hierarchy may be a reflection of the incidence of certain start-up types that also follows the urban hierarchy. To systematically account for the spatial context in which they operate, we distinguish between four region types: metropolitan regions, urban regions, rural regions and remote rural regions.

We present two main findings. First, there are only small differences across the urban hierarchy in terms of the labour market backgrounds of the entrepreneurs. This includes the start-up of spin-off firms by entrepreneurs with the same industry experience and start-ups out of necessity by entrepreneurs that used to be unemployed. Second, however, there is a large difference between urban and rural areas in terms of skill level present in the start-ups. This is manifested in two ways:

- Metropolitan and urban areas show significantly higher concentrations of entrepreneurship in knowledge-intensive business services.
- The education level of people involved in new firms is significantly higher in metropolitan and urban areas compared to their rural counterparts.

We show that the latter urban–rural divide in the education level of people involved in start-ups is not only an artefact of differences between urban and rural areas in terms of the industries in which start-ups operate. It is true also within industries. Start-ups in an industry in urban areas tend to employ more highly educated employees than start-ups operating in the same industry in rural areas. This gap appears as quantitatively important. Even in industries such as high-tech manufacturing and knowledge-intensive market services, the difference between urban and rural regions in terms of the fraction of university educated employees in start-ups amount to over 20 percentage points. In short, our results support the existence of an important skill gap between urban and rural areas in Sweden. This suggests that the sorting of human capital along the urban hierarchy can be a strong explanation of the differences in the economic impact of start-ups.

Why study the characteristics of entrepreneurship across regions?

Why study the characteristics of start-ups across the urban hierarchy? A simple answer is that new firms are different and have different prospects to grow, generate employment and influence the local economies they operate in. Ample studies have shown that entrepreneurs with more favourable skills and experiences outperform other entrepreneurs. Skills are difficult to measure directly, but the labour market careers of entrepreneurs contain information about the development of skills and experiences that can be appropriated in start-ups. As cases in point, studies have shown that entrepreneurs with previous entrepreneurship experience (Iversen *et al.*, 2010) and those coming from smaller firms (Lazear, 2004; Sørensen and Phillips, 2011) start relatively successful firms. Likewise, same industry experience (spin-offs) has proven to be a salient factor in explaining start-up performance. As a result of industry-specific skills and relevant experiences of the founders, spin-offs create more new jobs than other types of start-up (Klepper, 2002; Andersson and Klepper, 2013).

There are strong indications that there is indeed a spatial component in the characteristics of entrepreneurship. For instance, talent tends to sort itself into specific places in response to the unequal supply of labour market opportunities (Sjaastad, 1962; Broersma and Van Dijk, 2002). Human capital has a strong geographical dimension in that people with certain skills and expertise tend to locate in specific places (Moretti, 2012). As such, locational preferences may lead to differences in the regional distribution of successful entrepreneurs. Existing research on employee spin-off formation, for example, highlights the concentration of spin-offs, although at case-study level only. Employee spin-offs from one successful firm locate close to a parent firm and, in turn, churn out new spin-off firms in the same region (Klepper, 2007). This self-reinforcing cycle fosters the clustering of spin-off firms and the associated regional employment dynamics.

Studies that focus on the frequency of entrepreneurship but ignore the characteristics of new firms in different regions are thus likely to overlook an important dimension of spatial heterogeneity. In principle, two regions with similar start-up rates, i.e. pure frequency, may show vastly different performance if the nature of entrepreneurship in the two regions is different. Studies assessing the performance of start-ups do generally acknowledge different characteristics of the start-ups (see, for example, Acs, 2011). They are, however, mostly spatially blind in the sense that they have not considered the geography of the characteristics of such start-ups.

A number of recent studies on spatial differences in the employment effects of start-ups studies provide fuel to this line of argumentation. Fritsch and Mueller (2008), Van Stel and Suddle (2008) as well as Delfmann (2015) have established that the employment effects of entrepreneurship are larger in cities than they are elsewhere. Even though these studies speculate about the reasons why there may be spatial variation in the employment effect of start-ups, the sources of such spatial heterogeneity in the effects of entrepreneurship remain largely unexplained in the current literature. Fritsch and Schroeter (2011) develop and test more specific arguments without, however, explicitly embedding this in the conceptual framework considering the effect of entrepreneurship on employment generation. One reason why the effects of start-ups vary along the urban hierarchy is that the characteristics of the start-ups varies in parallel. Below, drawing on the work of Fritsch, we briefly expand on how the heterogeneity in start-ups fits in the current framework of employment effects of start-ups.

The regional employment effect of start-ups is twofold. There is a direct job effect, which are the jobs created in the start-ups (Fritsch and Noseleit, 2013). In addition, start-ups have indirect job effects because they exert competitive pressure on existing firms. The increased competition can cause other firms to shed jobs in the short run, but increased efficiency results in long-term positive job effects (Fritsch and Mueller, 2004; Koster, 2011; Koster and Stel, 2014). Also, through consumption effects, the employees and entrepreneurs of the start-ups sustain jobs in service firms.

Both direct and indirect job effects are constrained by space. Direct employment effects are clearly tied to the location of the start-up and are bound by the local labour market region. Also the indirect effects attenuate rapidly in space. Consumption effects are largely in untradeable products in the service industry (Graves and Linneman, 1979). In addition, spill-over effects resulting from cooperation, input–output relationships, knowledge sharing and competition are also mediated by proximity to other businesses (Breschi and Lissoni, 2001; Boschma, 2005). This implies that the employment effects in the regional economy are mediated by the characteristics of the regional economy. Spatial variations in the effect of entrepreneurship can be understood as differences in the direct employment effect (the employment dynamics in the new firms) and the indirect employment effect (the employment dynamics in incumbent firms).

The direct employment effects have been studied primarily at the firm level with the main research question which firms are most likely to be growing firms or high-impact firms (Acs, 2011; Fritsch and Schroeter, 2011). The prime approach to identify growing start-ups is to assess the characteristics of the founder or the founding team.

The likely spatial sorting of entrepreneurial talent as well as types of start-ups, does not however necessarily lead to increased direct employment effects in such places. First, it has been shown that start-ups in urban regions do tend to grow relatively quickly. Cities may provide the right incentives and inputs for growth or they may, as argued in the above, be the places where relatively successful entrepreneurs reside. At the same time, however, survival of start-ups is lower in cities, underlining the competitive environment in which they are active. The net direct employment effect of a cohort of start-ups may therefore be relatively small. Indeed, Fritsch and Noseleit (2013), as well as Andersson and Noseleit (2011), show that the net direct employment effect of start-ups is relatively small. Delfmann (2015), distinguishing between rural and urban areas, even finds that the net direct employment effect is larger in rural areas than in urban.

The employment effects of start-ups are not confined to job creation in the start-ups. The start-ups challenge existing firms that adapt to the newcomers; they change their strategies which implies indirect job effects (positive and negative) in the regional economy. These are the indirect employment effects. The literature has identified two important regularities in this respect. (1) The total employment effect of a cohort of start-ups takes between 8 and 12 years to materialize (Fritsch and Mueller, 2004; Koster, 2011): adaptation by others is not immediate. (2) The indirect employment effect is larger than the direct employment effect (Fritsch and Noseleit, 2013). As said, even though some start-ups grow, others remain stable or disappear relatively quickly. These effects cancel out after two years into the life of a cohort of start-ups; all remaining job dynamics are accounted for by the indirect effect. Also, the different employment effects between regions is fully accounted for by the indirect effect (Fritsch and Noseleit, 2013). There are two potential explanations. Again, the unequal distribution of types of start-up may explain the distinct indirect effects. Given the spatial sorting of human capital (Moretti, 2012), it is likely that certain regions attract successful entrepreneurs with growing businesses that have a wider influence on the regional economy. Even though this may not translate into employment effects in the firms themselves, they may trigger incumbent firms to react.

The regional context may influence the overall employment effect. If the context adapts well to the challenges set by the newcomers, the overall regional job effect may be larger. Fritsch and Noseleit (2013) suggest that the job effect in cities is larger because of the dense network that allows information to be passed on quickly. Also, the connection between the type of start-up and the regional context can be important. Industry-relatedness has been demonstrated to be an important driver of economic development (Neffke *et al.*, 2011) and this suggests that specific regions will benefit most from specific types of start-up. In line with such ideas, McCann and Ortega-Argilés (2013) argue that development trajectories are region-specific. Regions may thus not profit from entrepreneurship similarly. As shown, the nature of the start-ups involved may provide an important mechanism to explain the distinct direct and indirect employment effects.

Characterizing entrepreneurship across regions

Dimensions in the characteristics of entrepreneurship

The characteristics of entrepreneurship include a number of relevant dimensions (see, for example, Koster and Kapitsinis, 2015). One dimension is the pre-start-up situation of the entrepreneurs. A second approach is to look at the characteristics of the firm post-start-up. In the analysis, we will address both aspects. Concerning the first aspect, we distinguish between spin-offs, start-ups from unemployment, other start-ups and self-employed. We then characterize these types also in terms of industry membership and the human capital of the people involved in the new firm.

Types of new firms

We adopt the strategy of Eriksson and Kuhn (2006) as well as Andersson and Klepper (2013) and distinguish between types of start-up based on the previous labour market status of the initial set of employees in new firms. Specifically, we distinguish between four types of start-up:

1   self-employed
2   spin-offs
3   new firms by previously unemployed
4   other new firms.

Self-employed are all new firms with no employees, i.e. these are new firms which only involve one person. The other three categories of new firms are composed of sub-groups of new firms with two or more initial employees, i.e. new firms which have employees. Following Anderson and Klepper (2013), spin-offs are defined according to the backgrounds of their initial employees. If 50 per cent or more of the initial employees in a new firm worked at the same establishment in the previous year and constituted less than 50 per cent of the workforce at that establishment, we call this a spin-off. New firms founded by the previously unemployed are defined as all new firms with two or more initial employees, where all employees were unemployed in the previous year. Other new firms is a residual category consisting of all new firms with two or more initial employees that do not qualify as either spin-off or as a start-up by previously unemployed.

The main reason to distinguish between these four types of new firms is that previous research show that they differ vastly in terms of 'quality', as assessed by survival probabilities and post-entry employment growth. The analyses in, for example, Eriksson and Kuhn (2006) on Danish data and Andersson and Klepper (2013) on Swedish data show that employee spin-offs are a special class of new firms with lower exit hazard and higher employment growth at all ages. Spin-offs, as defined as above, outperform all other types of new firms even after controlling for factors such as initial size, industry affiliation and education level of the initial set of employees. New firms by previously unemployed people are likely to reflect necessity-based entrepreneurship where people are pushed into new ventures because of poor labour market prospects. Previous studies also show that this category of firms have substantially higher exit rates and lower levels of employment growth compared to other types of new firms. The category self-employed is interesting to separate from new firms with employees, not least since there are increasing arguments questioning using self-employment as a proxy for entrepreneurship (Henrekson and Sanandaji, 2014). It is also the case that in many countries as well as regions, new firms in the form of self-employed constitute the vast majority of the number of new firms. This means that variations in start-ups of new firms that have employees, such as spin-offs, are difficult to identify and quantify.

In summary, separating between the different types of start-up captures a pertinent dimension of the characteristics of entrepreneurship in regions that relate to the 'quality' and likely impact of start-ups in regions.

Industries

A typical argument in the literature on entrepreneurship is that innovative and knowledge-intensive new firms are likely to play a more important role for the economy compared to new firms in more traditional and less knowledge-intensive industries. One reason for this is that innovative start-ups are supposed to bring new ideas, new knowledge and new technologies to the market which stimulates economic renewal and increases variety in the economic system. This line of argument is akin to the Schumpeterian 'Mark I' hypothesis (Schumpeter, 1934), which emphasizes the role of small innovative firms as a driver of economic growth and structural transformation through 'creative destruction'.

The Schumpeterian innovation literature has for a long time recognized that the characteristics of the industry to which a firm (established and new) is affiliated can influence its innovation activity. Different industries have different technology and innovation opportunities and they are characterized by different technological regimes (Malerba and Orsenigo, 1993). Certain industries may be characterized by rapid technological progress, translating into high technology and innovation opportunities, which is typically the case in the early phases of a technology's life cycle (Vernon, 1966).

However, recent research on start-ups and high-growth firms suggests that there is no evidence that high-growth entrants are overrepresented in industries with large fractions of R&D, such as in high-tech manufacturing industries. If anything, high-growth entrants appear to be more frequent in knowledge-intensive services industries (Daunfeldt *et al.*, 2014, Henrekson and Johansson, 2010).

The industry is thus a relevant dimension, in terms of potential economic impact, on which start-ups can differ across places. Given the link between knowledge, innovation and potential economic impact, it is important to use an industry classification that accommodates differences in knowledge intensity rather than only differences in the goods or services produced. Eurostat's industry classification based on knowledge-intensity and technological advancement allows exactly this:[1]

1   high-tech manufacturing (HTM)
2   low-tech manufacturing (LTM)
3   high-tech knowledge-intensive services (HTKIS)
4   knowledge-intensive market services (MKIS)
5   knowledge-intensive financial services (KIFS)
6   other knowledge-intensive services (OKIS)
7   less knowledge-intensive services (LKIS)
8   other industries.

The classification involves a two-tier division. In the first step, it distinguishes between manufacturing and services industries. 'Other industries' forms the residual category and it comprises mainly entrants in agriculture, fishing and extraction industries. In the second step manufacturing and services are further disaggregated. High-tech manufacturing includes R&D and knowledge-intensive manufacturing industries such as pharmaceuticals, electronics, chemicals and aircraft. Low-tech manufacturing refers to more basic manufacturing such as rubber and plastic products, food products and furniture. A similar disaggregation is made for the services industries. High-tech knowledge-intensive services is computer programming, IT, R&D as well as broadcasting and music. Knowledge-intensive market services is in principle knowledge-intensive business services and comprise activities such as accounting, advertising and market research and architectural and engineering activities. Knowledge-intensive financial services is finance and insurance. Other knowledge-intensive services comprise activities such as veterinary, publishing, education and health. Less knowledge-intensive services is a broad category and consists for example of transport, real estate, wholesale and hotels.

Human capital

As argued earlier, there is plenty of evidence that human capital is a key factor to consider in explaining the performance and 'quality' of new firms. The so-called competence-based view of new firms (Colombo and Grilli, 2005), holds that the capabilities and qualities of new firms ultimately reside in the knowledge, skills and competences of the individuals involved in a new firm.[2]

A standard way to proxy complex matters like competences, knowledge and skills is to measure formal education. While far from perfect, this meas-ure has consistently been shown to capture relevant dimensions of human capital. For example, analyses of US and European regions show that highly educated workers are important drivers of regional development in terms of jobs, income per capita as well as entrepreneurship (Glaeser *et al.*, 1995, Cheshire and Magrini, 2000, Acs and Armington, 2004, Qian *et al.*, 2013).

Also at the level of the start-up, the human capital available to a firm is cru-cial for its success. People develop knowledge and skills, build up relevant pro-fessional networks and acquire problem-solving capabilities throughout the school system and their professional career. They can of course capitalize on the combined stock of experiences and accumulated knowledge by starting their own firm. Some studies show that successful business owners tend to rely on a wide variety of skills which allows them to oversee all aspects of the busi-ness, as formalized in the so-called 'jack-of-all-trades' theory (Lazear, 2004).

Still, even in explaining the success of new firms, relatively simple and straightforward measures like the fraction of employees in new firms that have a university degree tend to do a good job in capturing the role of human cap-ital. For example, Andersson and Klepper (2013) show in a detailed econo-metric analysis of the survival and growth of new firms in Sweden that the fraction of employees with a university degree has a statistically significant and highly robust influence on both survival and employment growth subse-quent entry. Conditional on an ample set of control variables, new firms with a higher fraction of initial employees with a university degree show higher survival as well as employment growth. One explanation for such a results is that formal education precisely proxies pertinent aspects of human capital, such as knowledge, skills and absorptive capacity. Better educated individuals may for example be in a better position to identify and create 'good' entrepreneurial opportunities.

We conclude that who are involved in new firms is a third relevant dimension of the nature of entrepreneurship. Here we measure human capital as the fraction of people involved in new firms that have a university education.

Data and empirical approach

We make use of Swedish data for the period 2008–2012. The underlying data is a matched employer–employee dataset maintained by Statistics Sweden. These data allow us to identify start-ups and categorize them according to their background characteristics using the so-called FAD-coding system.[3] Using this information we distinguish between self-employed, spin-offs, start-ups by unemployed and the residual category of 'other start-ups'. In addition, we have information on the employees involved in each start-up and the industry in which the firm is active. This allows us to distinguish between start-ups on the basis of industry and human capital according to the definitions presented earlier. All analyses pertain to the total number of start-ups in the period 2008–2012. As certain characteristics are rather rare this mitigates the issue of missing cases in the analysis. The total number of start-ups in the study period is 352,509. The corresponding employment figure is 447,851.

Locational information on each new firm allows us to explore the spatial heterogeneity that may exist in the characteristics of start-ups. Given that the found differences in the employment effect of start-ups follow the urban hierarchy (Fritsch and Schroeter, 2011), we use the Swedish urban hierarchy as the geographical context in the analysis. To this end, we employ a categorization of municipalities developed by the Swedish Board of Agriculture in which municipalities are categorized as belonging to a metropolitan region, urban region, countryside or remote countryside. Figure 8.1 maps the classification.[4]

Sweden has a strong urban–rural division with the most populous regions in the south and very scarcely populated regions in the north. The top of the urban hierarchy, the metropolitan regions, is relatively small (47 municipalities) and it consists only of municipalities belonging to the Stockholm, Göteborg and Malmö regions. Urban regions consist of medium-sized cities that serve the surrounding smaller municipalities. Countryside consists of municipalities without immediate access to a larger urban centre. Remote countryside are in essence all small municipalities in the northern part of Sweden located remotely, even from the larger centres in the north.

Table 8.1 presents basic descriptive information that serves as a backdrop for the analysis in the next section. By regional category, it reports the number of municipalities as well as the fraction of the Swedish population living

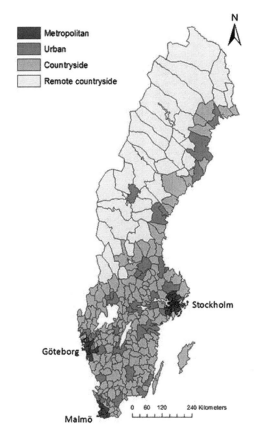

Figure 8.1 Categorization of Swedish municipalities

Table 8.1 Characteristics of four groups of municipalities in Sweden

Region classification	# Municipalities	Population	Employment	Emp. with university ed.
Metropolitan	47	36.5%	40.1%	48.6%
Urban	46	29.5%	29.6%	30.0%
Countryside	164	31.0%	27.5%	19.6%
Remote countryside	33	3.0%	2.8%	1.8%

there, its share in the overall employment and its share in the number of employees with a university education.

It is clear from the table that municipalities classified as metropolitan and urban host the vast majority of Sweden's population and employment. Metropolitan and urban municipalities together account for about 65 per cent of Sweden's population and about 70 per cent of total employment. At the same time they host almost 80 per cent of the country's workers with a university education. Interestingly, despite the Swedish system that promotes the decentralization of universities, there is a substantial overrepresentation of highly skilled employees in the metropolitan regions. The urban municipalities, in contrast, score in line with their proportions on the other indicators. This is in line with the current discussion regarding the spatial sorting of talent in a limited number of cities.

The countryside constitute the largest group of municipalities (164), but only account for about 30 per cent of population and employment. The remote countryside consist of a smaller number of municipalities in the north of Sweden with very limited population and employment size. They account for less than 3 per cent of Sweden's total population and employment, and less than 2 per cent of the country's employees with a university education.

Table 8.1 already suggests important sorting effects of talent. Whether this is actually reflected in the geography of the characteristics of entrepreneurship has not been studied, however. In line with Table 8.1, however, most of the knowledge-intensive activities in the country are concentrated in the bigger cities. Most of the internationally known technology-based start-ups tend to be geographically concentrated in Sweden's capital city, Stockholm. In fact, the international business press often describe Stockholm as a hub for technology-based start-ups.[5] More in general, Sweden is known as an advanced economy with a large high-tech industry. In most available international rankings, such as the World Bank's knowledge economy index or the EU innovation scoreboard, Sweden is consistently ranked as one of the most innovative and knowledge-intensive economies in the world (see e.g. Andersson *et al.*, 2013). In terms of entrepreneurship, Sweden is also often described as a 'hotbed' for start-ups in high-tech and knowledge-intensive industries. Success stories such as Skype, Spotify, King and Mojang provide good examples. Still, available studies that compare entrepreneurship in Sweden to other countries tend to conclude that the overall frequency of start-ups in Sweden in recent times does not appear markedly different in Sweden compared to elsewhere (Andersson and Klepper, 2013). Both arguments combined suggest that Sweden as a whole has relatively innovative and technologically advanced start-ups.

The characteristics of entrepreneurship across regions – stylized facts

In this section, we present an explorative analysis of the characteristics of start-ups along the urban hierarchy (Figure 8.1). As we are interested primarily in regional differences in the characterization of entrepreneurship, scale differences between the region types are abstracted away and we characterize entrepreneurship within each of the municipality groups. All percentages presented in Tables 8.2 through 8.5 sum to 100 per cent for each region type, save for some errors caused by rounding. We thus typify entrepreneurship in each region type and such an approach goes some way to explore the potential role of spatial sorting of entrepreneurship types.

Tables 8.2 and 8.3 show the geographical distribution of the different types of new firm defined on the basis of the background of the firm, the *ex-ante* characterization. Table 8.2 focuses on the number of start-ups, while Table 8.3 presents the same information for the total employment involved. Three main conclusions stand out.

First, and in line with previous studies, it is clear that self-employment is by far the most prominent start-up mode. In total, over 85 per cent of all new firms involve only one person. Given that at least the immediate direct employment effects of such firms are small, this again suggests the relevance of distinguishing between different types of start-ups in regional analyses of the employment effect of start-ups. Most start-ups are small and may not have a lasting impact on the regional economy.

Second, Table 8.2 reveals that the prevalence of the start-up types varies only slightly across the four regional regimes. Although distinguishing between start-up types along the lines of the background of the entrepreneur may be relevant, it does not, at face value, seem to be a strong explanation for regional differences in the impact of entrepreneurship. There is no clear suggestion of spatial sorting in terms of self-employment or the labour market background of the founders. This is further substantiated by the fact that the diffuse 'other start-up' group is relatively large, particularly in terms of employment. Given the small sizes of the spin-off and unemployment groups they are not likely to have a substantial economic impact, even though spin-offs have been shown to be successful start-ups relative to other firms. In other words, known labour market routes towards successful entrepreneurship, such as having industry experience, are important in understanding firm performance. Such routes, rare as they are, may carry little weight in explaining regional differences in the effect of start-ups.

Third, we can deduct that start-ups in the metropolitan regions are relatively large. In Table 8.3, which addresses employment, the shares of self-employed have gone down compared to Table 8.2. By definition, start-ups in the form of self-employment only have one employee whilst the other categories are larger firms. This automatically reduces the shares of self-employment expressed employment. Interestingly, this effect is strongest for metropolitan regions. In terms of establishments there is a 4.8 per cent difference between metropolitan and remote rural regions in the share of self-employed. If the difference is expressed in employment share, it all but doubles to 9.3 per cent. This means that the other start-ups in metropolitan regions are relatively large. This underlines the potential relevance of human capital in any explanation of

Table 8.2 Share of start-up types by region, number of start-ups (N=352,509)

	Self-employed	Spin-offs	Unemployed	Other
Metropolitan	84.1%	1.2%	2.5%	12.2%
Urban	85.4%	1.2%	2.1%	11.3%
Countryside	88.2%	0.9%	1.7%	9.2%
Remote countryside	88.9%	0.8%	1.2%	9.0%
Total	85.8%	1.1%	2.1%	10.9%

Table 8.3 Share of start-up types by region, employment in start-ups (N=447,851)

	Self-employed	Spin-offs	Unemployed	Other
Metropolitan	64.2%	3.9%	4.1%	27.8%
Urban	66.6%	3.9%	3.5%	26.1%
Countryside	72.6%	3.1%	3.0%	21.3%
Remote countryside	73.5%	2.7%	2.2%	21.6%
Total	67.6%	3.6%	3.5%	25.3%

spatial heterogeneous employment impact of start-ups. Telling in this respect is that the difference between start-up numbers and employment share are stable across the types of start-up. In other words, metropolitan start-ups are larger than rural start-ups independent of the type of firm. As an aside, we see that spin-offs trump the other types of start-up in terms of employment at start-up. This corroborates the stylized fact that spin-off firms are relatively successful start-ups.

Tables 8.4 through 8.7 present the results regarding the organizational characteristics of the firm, the *ex-post* characterization. First, we turn to the industry breakdown of the start-ups, in terms of number of firms (Table 8.4) and employment (Table 8.5). Again, the data confirm well-known patterns in the sense that by far the biggest group of start-ups in the Swedish economy, 63.7 per cent, is in business services. The variation across regional regimes is however much larger than when looking at the types of start-ups in terms of labour market background. In metropolitan regions, start-ups are predominantly in the knowledge-intensive service industries (45.8 per cent in total), which is a much higher share than in any of the other regions. Indeed, the corresponding figure for remote rural regions is only 13.3 per cent (total). The shares neatly follow the urban hierarchy with urban and rural regions taking up the middle positions. The dominancy of the three metropolitan regions for KIS start-ups is underlined by the fact that out of all KIS start-ups in Sweden 58.6 per cent are located in one of the metropolitan areas. This is strongly suggestive of the crucial role of human capital as the driver of geographical differences in the nature, and the economic impact, of start-ups. In line with this interpretation is the much flatter distribution across the regional regimes of the services that are less knowledge-intensive (LKIS). The spread for LKIS

Table 8.4 Industry distribution of start-ups by region, number of start-ups (N=352,509)

Sector	Metropolitan	Urban	Countryside	Remote countryside	Total
HT man	0.4%	0.5%	0.5%	0.4%	0.5%
LT man	2.5%	3.4%	4.4%	4.6%	3.4%
HTKIS	8.0%	4.7%	2.3%	1.5%	5.2%
MKIS	23.6%	16.1%	9.9%	6.7%	17.0%
KIFS	0.9%	0.5%	0.3%	0.2%	0.6%
OKIS	13.3%	9.1%	6.4%	5.0%	9.9%
LKIS	32.2%	33.0%	29.0%	24.5%	31.1%
OTHER	19.1%	32.8%	47.2%	57.3%	32.5%
TOTAL	100%	100%	100%	100%	100%

Note: HT man and LT man stands for high-tech and low-tech manufacturing. HTKIS, MKIS and KIFS refer to high-tech knowledge intensive services, knowledge-intensive market services and knowledge-intensive financial services. OKIS is other knowledge-intensive services. LKIS is less knowledge-intensive services and OTHER is uncategorized sectors.

Table 8.5 Industry distribution of start-ups by region, in employment (N=447,851)

Sector	Metropolitan	Urban	Countryside	Remote countryside	Total
HTM	0.4%	0.5%	0.6%	0.4%	0.5%
LTM	2.4%	3.6%	4.7%	5.1%	3.5%
HTKIS	7.6%	4.5%	2.1%	1.3%	5.0%
MKIS	21.8%	15.1%	9.6%	6.7%	16.1%
KIFS	1.0%	0.5%	0.3%	0.1%	0.6%
OKIS	12.3%	8.9%	6.6%	5.2%	9.5%
LKIS	36.4	37.1%	33.0%	29.1%	35.3%
OTHER	18.1%	29.9%	43.2%	52.0%	29.6%
TOTAL	100%	100%	100%	100%	100%

Note: HTM and LTM stands for high-tech and low-tech manufacturing. HTKIS, MKIS and KIFS refer to high-tech knowledge-intensive services, knowledge-intensive market services and knowledge-intensive financial services. OKIS is other knowledge-intensive services. LKIS is less knowledge-intensive services and OTHER is uncategorized sectors.

is only 7.7 per cent. The underlying spatial distribution of human capital is thus reflected in the industries in which start-ups are active. Interestingly, though only constituting a small share of all start-ups, high-tech manufacturing start-ups also show a flat distribution across the urban hierarchy. Even though innovation and human capital is crucial for such industries it is likely less central to the production process making such start-ups less responsive to the benefits from the urban/metropolitan environment.

Table 8.6 Share of employees with a university education by start-up type by region

Region	Self-employed	Spin-offs	Unemployed	Other	All start-ups
Metropolitan	48.7%	35.5%	33.6%	31.8%	42.9%
Urban	36.4%	23.1%	25.9%	22.7%	32.0%
Countryside	25.8%	13.8%	22.3%	17.3%	23.5%
Remote countryside	21.2%	13.0%	17.2%	14.7%	19.5%

The human capital aspect is addressed more directly in the final set of tables (Tables 8.6 and 8.7). Table 8.6 shows the share of employees with a university education for each of the start-up types, as well as for the total group of start-ups by region. Table 8.7 shows the same information, but for the industry breakdown. The results reiterate the sorting of talent towards urban areas. Table 8.6 shows that in metropolitan areas, for all start-ups taken together, 42.9 per cent of all employees (owners included) have a university education. The corresponding figure for remote regions is 19.5 per cent only. Again, the shares precisely follow the urban hierarchy. Also the comparison across start-up groups is relevant and it shows that self-employed are relatively highly educated, in all regions, and that the other groups have a lower input of human capital. The overrepresentation of highly educated self-employed is highest in metropolitan regions and this may be a reflection of the under-lying skill distribution across space: metropolitan areas harbour more highly educated potential entrepreneurs. It seems that self-employment is the pre-ferred mode of appropriating the skills available. Spin-offs, for example, score much lower in terms of highly-skilled employment. This suggests a role of the economic context in stimulating highly skilled people to take the step towards self-employment. Finally, we concluded earlier that the diffuse group of 'other start-ups' may deserve further scrutiny given the relatively large size of this group. It, however, scores lowest in terms of highly educated employment.

The high shares of employees with a university degree in metropolitan areas may be the result of the specific industry structure in those places. Table 8.5 indeed shows that certain knowledge-intensive industries are concentrated in metropolitan regions. The results in Table 8.6 may be a reflection of this. More technically, Table 8.5 and 8.6 may measure the same underlying phenomenon. As the final step of the analysis, and to address this, Table 8.7 gives the share of highly educated personnel by industry by region type. The results recon-firm the skill gap between metropolitan and rural regions, particularly for the knowledge intensive services (HTKIS, MKIS, KIFS, OKIS). The shares for metropolitan and rural regions differ by as much as 20 percentage points within the same industry. Again the urban hierarchy is followed and it may thus be more appropriate to speak of a skill slope down the urban hierarchy. The results are strongly indicative of the heterogeneity or functional special-ization within industries. The different skill shares for manufacturing, for

Table 8.7 Share of employees with a university education by industry

Region\sector	HTM	LTM	HTKIS	MKIS	KIFS	OKIS	LKIS	OTHER	Total
Metropolitan	40.8%	30.1%	65.0%	62.5%	60.8%	63.9%	28.0%	27.2%	42.9%
Urban	32.5%	21.8%	62.8%	56.6%	49.7%	56.7%	21.4%	21.7%	32.0%
Countryside	21.7%	17.5%	51.9%	48.1%	40.3%	49.2%	18.2%	17.3%	23.5%
Remote countryside	23.2%	16.3%	44.7%	43.4%	12.5%	44.5%	15.9%	15.5%	19.5%

Note: HTM and LTM stands for high-tech and low-tech manufacturing respectively. HTKIS, MKIS and KIFS refer to high-tech knowledge-intensive services, knowledge-intensive market services and knowledge-intensive financial services respectively. OKIS is other knowledge-intensive services. LKIS is less knowledge-intensive services and OTHER is uncategorized sectors.

example, suggest spatial heterogeneity in the activities performed within the same broadly defined industry. For knowledge-intensive services it becomes clear that they are much more knowledge-intensive in city regions than in other parts of the country. These examples also show the shortcomings of measuring knowledge intensity of regions on the basis of industry classifications. It appears that the skill gap between regions may then be underestimated as the difference in high-skilled employees is larger than differences in industry structure.

The skill gap between the regions is also present for the industries with a lower knowledge input (LTM, LKIS and OTHER), but it much less pronounced. They appear to have a lower need, or possibility, to capitalize on the knowledge benefits that cities may offer.

Summary and conclusion

The literature on the geography of start-ups has for a long time primarily focused on analysing the frequency of entrepreneurship and neglected how local conditions may influence the characteristics of start-ups along dimensions such as the industries in which new firms operate, types of new firms and background of entrepreneurs and their employees. While the determinants of the overall rate of start-ups is well understood, there is a gap in our knowledge of how the composition and characteristics of new firm formation vary spatially. This gap has become more evident in view of recent analyses documenting that employment effects of start-ups vary distinctively along urban–rural hierarchy. The effect start-ups have on regional employment growth is significantly higher in urban compared to rural areas. One reason for this discrepancy could be that start-ups differ in their characteristics along the urban–rural hierarchy, which calls for studies on spatial variations in start-up characteristics.

Against this backdrop, this chapter has explored spatial variations in the characteristics of start-ups with a focus on variations along the urban–rural hierarchy. We focused on characteristics in terms of (1) types of start-ups, (2) the industries in which they operate and (3) the level of education of the people involved in the new firms. Based on an assessment on data for Sweden over the period 2008–2012 our main findings are:

1   The composition of start-ups in terms of the types of new firms that are started does not vary markedly along the urban–rural hierarchy. For example, there are no big differences between urban areas and the countryside in terms of the fraction of new firms that are employee spin-offs or new firms started by previous unemployed people.

2   Urban and rural areas differ primarily with respect to the knowledge-intensity of start-ups. This is manifested by (1) urban areas showing significantly higher concentrations of entrepreneurship in knowledge-intensive business services; and (2) the education level of

people involved in new firms being significantly higher in urban areas compared to their rural counterparts. Start-ups in an industry in urban areas tend to employ more highly educated employees than start-ups operating in the same industry in rural areas. Even in industries such as high-tech manufacturing and knowledge-intensive market services, the difference between urban and rural regions in terms of the fraction of university educated employees in start-ups amount to over 20 percentage points.

One interpretation of these findings is that they reflect an underlying difference in knowledge resources between urban and rural areas, leading to a 'modernity gap' in the characteristics of start-ups. It is a natural conjecture that the characteristics of start-ups adjust to the underlying local resource base, in particular the supply of skills. This means in turn that spatial variations in the characteristics of entrepreneurship depend on how knowledge resources, skills and talent sort in geography.

The urban–rural skills gap as reflected in the urban–rural divide in characteristics of start-ups could very well be understood as a consequence of sorting to the bigger cities. The spatial distribution of skills and talent is indeed far from random. Ample studies of spatial sorting of workers show that more skilled and more educated workers sort themselves to bigger cities and urban regions (Glaeser and Maré, 1994; Combes *et al.*, 2008; Andersson *et al.*, 2014). We showed in this chapter that almost 50 per cent of the employees in the metropolitan regions of Sweden have a university education while the corresponding figure for the countryside is only about 20 per cent. This clearly gives better conditions for knowledge-intensive entrepreneurship in the bigger cities. In fact, such places could have disproportional economic benefits from entrepreneurship because of the relatively high 'quality' of the entrepreneurs and employees of new firms. Such an argument tallies with the well-developed literature on the geography of talent (Moretti, 2012). Previous research show that more knowledge-intensive start-ups with better prospects to survive generate greater employment effects locally, and the urban–rural differences in the characteristics of start-ups documented in this chapter could be one reason for the urban–rural discrepancy in the effect of start-ups on local growth.

In summary, this study shows that acknowledging the heterogeneity of start-ups is crucial in further exploring and understanding the geography and economic impact of entrepreneurship. The chapter documents in this regard the crucial role of human capital in understanding spatial variations in the characteristics of start-ups, thus lining up with more general studies on the geography of talent. Even though the results are suggestive of spatial sorting of successful entrepreneurs, further studies should more formally address the relationships between the characteristics of entrepreneurship, skills and knowledge resources, and employment generation. Can we better understand the role that the spatial sorting of skills plays in influencing spatial variations in the characteristics of entrepreneurship? How important is the local supply of skills? Can we develop an explanatory empirical model to try to explain spatial differences in the composition of start-ups? Are the differences in the characteristics of entrepreneurship between urban and rural areas responsible for the urban–rural divide in the effects of entrepreneurship on local growth? Are the direct or indirect employment effects more prominent? And is the local environment, in addition to sorting, also important as a conducer of the employment effect? Finally, it should also be noted that this chapter used data for Sweden which is an advanced OECD economy with a relatively large high-tech and knowledge-intensive industry. Analyses for other countries with different economic structures would be interesting.

Notes

1 This classification builds directly on Eurostat's definitions available at http://ec.europa.eu/eurostat/cache/metadata/Annexes/htec_esms_an3.pdf.
2 For example, Cooper and Bruno (1977, p. 21) state that 'any competitive advantage the new firm achieves is likely to be based upon what the founders can do better than others'. While the founders certainly have a special role, the same argument could also apply to the whole set of individuals involved in new firms, i.e. including not only founders but also employees.
3 FAD (Företagens och Arbetsställens Dynamik) is a coding scheme for establishments and firms which distinguish various types of new firms based on worker flows.
4 Several different criteria are used to categorize municipalities, including in- and out-commuting as a fraction of total employment, percent of populated surface and distance to large city (Swedish Agricultural Board, Report 2009:2).
5 See www.telegraph.co.uk/finance/newsbysector/mediatechnologyandtelecoms/11689464/How-Sweden-became-the-startup-capital-of-Europe.html.

References

Acs, Z. (2011). High-impact firms: gazelles revisited. In Fritsch, M. (ed.), *Handbook of Research on Entrepreneurship and Regional Development*. Cheltenham, UK and Northampton, MA: Edward Elgar.

Acs, Z. and Armington, C. (2004). Employment growth and entrepreneurial activity in cities. *Regional Studies*, 38(8), 911–927.

Andersson, M. and Klepper, S. (2013). Characteristics and performance of new firms and spinoffs in Sweden. *Industrial and Corporate Change*, 22(1), 245–280.

Andersson, M. and Noseleit, F. (2011). Start-ups and employment dynamics within and across sectors. *Small Business Economics*, 36(4), 461–483.

Andersson, M., Klaesson, J. and Larsson, J.P. (2014). The sources of the urban wage premium by worker skills – spatial sorting or agglomeration economies? *Papers in Regional Science*, 93(4), 727–747.

Andersson, M., Anokhin, S., Autio, E., Ejermo, O., Lavesson, N., Lööf, H., Savin, M., Wincent, J. and Ylinenpää, H. (eds) (2013). *Det Innovativa Sverige – Sverige som kunskapsnation i en internationell kontext*. Stockholm: ESBRI och Brunzell Design.

Boschma, R. (2005). Proximity and innovation: a critical assessment. *Regional Studies*, 39(1), 61–74.

Breschi, S. and Lissoni, F. (2001). Knowledge spillovers and local innovation systems: a critical survey. *Industrial and Corporate Change*, 10(4), 975–1005.

Broersma, L. and van Dijk, J. (2002). Regional labour market dynamics in the Netherlands. *Papers in Regional Science*, 81(3), 343–364.

Cheshire, P. and Magrini, S. (2000). Endogenous processes in European regional growth: convergence and policy. *Growth and Change*, 31(4), 455–479.

Colombo, M.G. and Grilli, L. (2005). Founders' human capital and the growth of new technology-based firms: a competence-based view. *Research Policy*, 34(6), 795–816.

Combes, P.P., Duranton, G. and Gobillon, L. (2008). Spatial wage disparities: sorting matters! *Journal of Urban Economics*, 63(2), 723–742.

Cooper, A.C. and Bruno, A.V. (1977). Success among high-technology firms. *Business Horizons*, 20(2), 16–22.

Daunfeldt, S.O., Elert, N. and Johansson, D. (2014). Are high-growth firms overrepresented in high-tech industries? HUI working papers no. 103.

Delfmann, H. (2015). Understanding entrepreneurship in the local context: population decline, ageing and density. University of Groningen, PhD thesis.

Eriksson, T. and Kuhn, J.M. (2006). Firm spin-offs in Denmark 1981–2000: patterns of entry and exit. *International Journal of Industrial Organization*, 24(5), 1021–1040.

Fritsch, M. (2013). New business formation and regional development: a survey and assessment of the evidence. *Foundations and Trends in Entrepreneurship*, 9(3), 249–364.

Fritsch, M. and Mueller, P. (2004). Effects of new business formation on regional development over time. *Regional Studies*, 38(8), 961–975.

Fritsch, M. and Mueller, P. (2008). The effect of new business formation on regional development over time: the case of Germany. *Small Business Economics*, 30(1), 15–29.

Fritsch, M. and Noseleit, F. (2013). Investigating the anatomy of the employment effect of new business formation. *Cambridge Journal of Economics*, 37(2), 349–377.

Fritsch, M. and Schroeter, A. (2011). Why does the effect of new business formation differ across regions? *Small Business Economics*, 36(4), 383–400.

Glaeser, E.L. and Maré, D.C. (1994). Cities and skills. *Journal of Labor Economics*, 19, 316–342.

Glaeser, E.L., Scheinkman, J. and Shleifer, A. (1995). Economic growth in a cross-section of cities. *Journal of Monetary Economics*, 36(1), 117–143.

Graves, P.E. and Linneman, P.D. (1979). Household migration: theoretical and empirical results. *Journal of Urban Economics*, 6(3), 383–404.

Henrekson, M. and Johansson, D. (2010). Gazelles as job creators: a survey and interpretation of the evidence. *Small Business Economics*, 35(2), 227–244.

Henrekson, M. and Sanandaji, T. (2014). Small business activity does not measure entrepreneurship. *Proceedings of the National Academy of Sciences*, 111(5), 1760–1765.

Iversen, J., Malchow-Møller, N. and Sørensen, A. (2010). Returns to schooling in self-employment. *Economics Letters*, 109(3), 179–182.

Klepper, S. (2002). The capabilities of new firms and the evolution of the US automobile industry. *Industrial and Corporate Change*, 11(4), 645–666.

Klepper, S. (2007). Disagreements, spinoffs, and the evolution of Detroit as the capital of the US automobile industry. *Management Science*, 53(4), 616–631.

Koster, S. (2011). Individual foundings and organizational foundings: their effect on employment growth in the Netherlands. *Small Business Economics*, 36(4), 485–501.

Koster, S. and Kapitsinis, N. (2015). Analysing the geography of high-impact entrepreneurship. In Karlsson, C., Andersson, M. and Norman, T. (eds),

Handbook of Research Methods and Applications in Economic Geography.
Cheltenham: Edward Elgar.

Koster, S. and Stel, A. (2014). The relationship between start-ups, market mobility and employment growth: an empirical analysis for Dutch regions. *Papers in Regional Science*, 93(1), 203–217.

Lazear, E.P. (2004). Balanced skills and entrepreneurship. *American Economic Review*, 208–211.

McCann, P. and Ortega-Argilés, R. (2013). Smart specialization, regional growth and applications to European Union cohesion policy. *Regional Studies*, 49(8), 1291–1302.

Malerba, F. and Orsenigo, L. (1993). Technological regimes and firm behavior. *Industrial and Corporate Change*, 2(1), 45–71.

Moretti, E. (2012). *The New Geography of Jobs.* Boston and New York: Houghton Mifflin Harcourt.

Neffke, F., Henning, M. and Boschma, R. (2011). How do regions diversify over time? Industry relatedness and the development of new growth paths in regions. *Economic Geography*, 87(3), 237–265.

Qian, H., Acs, Z. and Stough, R.R. (2013). Regional systems of entrepreneurship: the nexus of human capital, knowledge and new firm formation. *Journal of Economic Geography*, 13(4), 559–587.

Schumpeter, J.A. (1934). *The Theory of Economic Development: An Inquiry into Profits, Capital, Credit, Interest, and the Business Cycle.* London: Transaction publishers.

Sjaastad, L.A. (1962). The costs and returns of human migration. *The Journal of Political Economy*, 80–93.

Sørensen, J.B. and Phillips, D.J. (2011). Competence and commitment: employer size and entrepreneurial endurance. *Industrial and Corporate Change*, 20(5), 1277–1304.

Van Stel, A. and Suddle, K. (2008). The impact of new firm formation on regional development in the Netherlands. *Small Business Economics*, 30(1), 31–47.

Vernon, R. (1966). International investment and international trade in the product cycle. *The Quarterly Journal of Economics*, 80(2), 190–207.

9 Entrepreneurship in peripheral regions

A relational perspective

Sandra Bürcher, Antoine Habersetzer and Heike Mayer

Introduction

Entrepreneurship is often seen as a major driver of economic change (Malecki, 1994; Metcalfe, 2004) and regional development dynamics – particularly in peripheral regions – are shaped by the ways in which individual entrepreneurs discover, evaluate and exploit business opportunities (Shane, 2003). The importance of investigating entrepreneurial behavior from a geographic perspective is understandable when bearing in mind that entrepreneurial activity tends to spatially cluster, and is thus very unevenly distributed in space. In recent years, economic geographers have started to examine regional disparities in entrepreneurship (Audretsch, Falck, Feldman & Heblich, 2012; Bosma & Schutjens, 2011). However, these studies often focus on core regions and many insights about entrepreneurial behavior are derived from the urban context. Less prominent are studies of entrepreneurial dynamics in peripheral regions (Baumgartner, Pütz & Seidl, 2013; Benneworth, 2004; Freire-Gibb & Nielsen, 2014; Vaillant & Lafuente, 2007). Moreover, regional disparities in entrepreneurial behavior are rarely examined when talking about differences *between* peripheral areas. Such a perspective recognizes the heterogeneity of peripheral regions and might help us understand why certain peripheral regions are more successful than others.

In the context of peripheral areas, several structural and relational factors hamper entrepreneurial agency and thus reduce the competitiveness of peripheral regions. However, differences between peripheral regions are very accentuated (OECD, 2006) and highly entrepreneurial and globally competitive firms can also be found in peripheral regions (Simon, 2009). Yet, insights why some peripheral regions show successful entrepreneurship, while other similarly disadvantaged and peripherally located regions do not, are sparse and more research is needed to understand the drivers and barriers of entrepreneurship in the periphery.

In this chapter, we review determinants of divergent entrepreneurial dynamics in peripheral regions by employing a relational perspective. We argue that successful entrepreneurship in peripheral regions relies on two different types of relations: on the one hand, economic relations in a strict sense, consisting of relations firms form in order to acquire relevant business knowledge; and on the other hand, economic relations in a wider sense, consisting of relations entrepreneurs form in order to shape regional contexts. The latter is called regional engagement of entrepreneurs and has direct or indirect consequences for the competitiveness of the region.

In the following we discuss a number of conceptual issues that are relevant when exploring entrepreneurship in peripheral regions. We concentrate on specificities of entrepreneurship in the periphery and the usefulness of a relational perspective. We then discuss in more detail a conceptual approach regarding the differentiation between the above-mentioned types of firm relations. Specifically, we present two approaches that are rooted in relational economic geography – namely the entrepreneurial heritage and the embeddedness approach – which may allow for a more sophisticated understanding of entrepreneurial dynamics in the periphery. We explain why these are specifically important for firms and entrepreneurs in peripheral areas, and why it makes sense to discuss them together. With this enlarged perspective, we aim to give a conceptual framework for empirical research explaining the presence of successful entrepreneurial firms in, and differences between, peripheral areas. The conceptual framework also encompasses a typology of four kinds of regional peripheral economies based on different characteristics concerning their entrepreneurial heritage and spatial embeddedness. On the basis of this typology we deducted some general policy recommendations. The final section summarizes the conceptual arguments, discusses possible contributions to relational economic geography and implications for further research.

Entrepreneurship in the periphery

When discussing entrepreneurship in the periphery, Lagendijk and Lorentzen's (2007) definition of peripheries as regional economies located outside of metropolitan areas is useful. The distance or the proximity of a firm's location to core regions is generally seen as one important distinguishing feature. Besides the purely spatial, Torre and Gilly (2000) also emphasize the importance of organizational proximity of firms. Taking a rather structural approach, the term geographical proximity is based on spatial distance, and the social processes modifying natural constraints on mobility. On the other hand, "interaction between actors and the modalities of co-ordination" (Torre & Gilly, 2000, p. 174) are the determining factors for identifying organizational proximity between actors. It is important to emphasize the high heterogeneity of firms in peripheral areas: While some firms might be highly competitive and indeed organizationally proximate to firms in core regions, the majority of firms in peripheral regions are much less competitive and thus organizationally distant from firms in core regions. Consequently, these two types of peripheral firms are organizationally distant from each other, even though they might be geographically quite proximate. The analysis of the composition of a regional firm population will thus not only reveal the differences with firm populations of core areas, but also heterogeneities between different peripheral firm populations.

In this sense, the varying degree of interaction and coordination both within a regional economy and within larger, often global networks can be seen as a pivotal element for the disparities in innovativeness between regional firm populations, and for categorizing regional economies as central or peripheral. Consequently, a regional economy can be geographically proximate, but organizationally distant from core regions. Thus, reducing the definition of peripherality to spatiotemporal distance would not live up to the ordering principles of the modern globalized knowledge economy, characterized by fragmented economic spaces and an ongoing concentration of wealth and power in increasingly interconnected core regions (Anderson, 2000). The peripherality of non-metropolitan areas is also due to their limited capacity to connect to these networks of globalized economic places, or because of their lower hierarchical position within these networks (Lagendijk & Lorentzen, 2007). The peripherality of a certain regional economy not only concerns economic aspects, but also aspects of institutional decision-making processes as the decision-makers are often located in core regions at different levels (national and supranational).

Often defined as the opposite of urban areas, the periphery is generally considered to be "cast in a residual role" (Ward & Brown, 2009, p. 1238). Because of low production costs, especially cheap land and the availability of unskilled labor, the only significant industrial activity presumed in peripheral regions are branch plants from large companies of mature industries, whose headquarters and innovation activities are located in the core (Boschma & Lambooy, 1999). The innovativeness of firms is often related to the endogenous development potential of regional economies, and it is common to ascribe a lack of innovativeness and technological dynamism to peripheral regions (Copus, Skuras & Tsegenidi, 2008). A frequently used framework to explore the innovativeness of regional economies is the Regional Innovation Systems (RIS) approach, which is also applied in the context of peripheral regions. The most common deficiencies of RIS in peripheral regions are subsumed under the terms "organizational thinness" and "institutional thinness," which are characterized by four main properties: First, the predominance of small and medium sized enterprises and branch plants active in mature industries, which in turn have minor R&D activities and a poorly qualified workforce. As a result, this limits the absorptive capacity of local firms and reduces the possibility of achieving radical innovations. Second, the relatively small population of firms is often not sufficient to initiate a self-reinforcing clustering process and to generate significant agglomeration economies. Third, important support organizations, like universities and specialized services, as well as formal and informal institutions are absent or weakly developed. And fourth, firms show a lower degree of network connectedness, which reduces the possibility of knowledge diffusion (Isaksen, 2015; Tödtling & Trippl, 2005).

Notwithstanding, peripheral regions show a high diversity of development patterns (OECD, 2006). As a matter of fact, very entrepreneurial and innovative firms, and even global market leaders, can be found in peripheral regions (Simon, 2009). Examples of globally competitive industries in peripheral regions are, among others, the watch-making industry in the Swiss Jura Arc (Maillat, Lecoq, Nemeti & Pfister, 1995), the ICT industry in the Oulu region, Finland (Virkkala, 2007) and the metallurgical complex in Lister, Norway (Isaksen, 2015). Thus, more attention should be paid to differences between peripheral regions (Meccheri & Pelloni, 2006), or more precisely to different types of firms and different types of regional engagement within these heterogeneous peripheral areas.

The relational perspective seems to offer great potential to explain differences in performance between peripheral regions. On the one hand, relational perspectives are increasingly applied to explore knowledge acquisition strategies of firms. Since "firms act in relational spaces rather than anonymous market spaces" (Schutjens & Stam, 2003, p. 115), it is not sufficient to simply discuss the aggregated specificities of a regional economy. On the other hand, networks can be understood as an important basis for collective action. By this, economic actors can shape regional economies more effectively.

Firms are embedded in various types of networks, with different repercussions on their economic performance. We broadly distinguish two different categories of economic relations: first, economic relations in a strict sense, consisting of relations between firms with the function of acquiring relevant business knowledge; and second, economic relations in a wider sense, consisting of relations of entrepreneurs with the function of shaping regional, e.g. institutional or organizational, contexts. While they describe very different entrepreneurial activities, they both have important repercussion for a firm's competitiveness. Even more, they are especially important for firms located in peripheral areas: Because of the specificities of peripheral locations, firms are challenged to acquire external knowledge and collectively advocate for a better economic framework. Thus, engaging in the two types of networks is crucial for entrepreneurial firms in peripheral areas to remain competitive. Further, we assume that the capacity to successfully connect to and act within both types of networks is based on a rather similar set of competences. Firms that are better integrated in knowledge networks are supposedly also more active in engagement networks that help shape regional contexts. Due to this co-evolution of knowledge networks and engagement networks, economic relations in a strict and a wider sense have to be studied together.

Since the regional economies of peripheral areas are characterized by a number of deficiencies concerning local knowledge spillovers and "local buzz" (Grillitsch & Nilsson, 2015), firms especially rely on external relations, so-called "global pipelines" (Bathelt, Malmberg & Maskell, 2004). A high outbound connectedness for firms located in the periphery is vital to acquire relevant information on market dynamics and to assure that the decision-makers in the core are sensitized concerning the economically challenging contexts in peripheral regions. Empirical research conducted in several countries shows that firms in peripheral areas compensate organizational thinness and limited knowledge spillovers by strengthening their extra-regional relations and by connecting to extra-regional knowledge sources (Doloreux, 2003; Rodríguez-Pose & Fitjar, 2013). However the capabilities to do so strongly differ between firms, depending on their absorptive capacity, local knowledge diffusion mechanisms (Cabiddu & Pettinao, 2013; Grillitsch & Nilsson, 2015) and their spatial embeddedness (Oinas, 1997), i.e. the spatial distribution of their social capital.

Moreover, peripheral regions are characterized by stronger informal networks (for a list, see Atterton, 2007). Such informal networks can act as a source of support and knowledge compensating the absence of more formal information sources. When doing business in peripheral areas, face-to-face contacts seem to be crucial (Atterton, 2007). Such dense social networks can lead to a higher level of trust and a reduced risk of opportunistic behavior, thus increasing the propensity for, and quality of, knowledge exchange (Grillitsch & Nilsson, 2015) and reducing the risks related to entrepreneurial actions (Westlund & Bolton, 2003). Also, these dense networks can engender an increased sensitivity for common regional interests and a more efficient interest articulation vis-à-vis various institutions or other regional actor groups.

Social relations that are too strong and high levels of reciprocal control can also hamper entrepreneurial initiatives (Grabher, 1993) and lead to so-called over-embeddedness (Uzzi, 1996). Atterton (2007), for example, examines the different degrees of strength of ties between business owners by analyzing three towns in the highly peripheral Highlands and Islands of Scotland situated at different distances from the region's main market center, Inverness. She found that there are differences between the three towns she had explored concerning, for example, the strength of the ties. She also points out that networks that are too strong constitute the risk of over-embeddedness and therefore lock-in. In this context it is important to distinguish between strong and weak ties (Granovetter, 1973) or as Putnam (2000) calls it, bonding and bridging social capital. Granovetter (1973, p. 1378) argues that weak ties are "indispensable to individuals' opportunities." They are therefore of utmost interest in the context of peripheral regions as they can balance the risk of over-embeddedness in strong ties. Weak ties might explain differences between peripheral economies, as they seem to be indispensable for firms to be competitive.

Three stylized facts can be summarized from the review so far: First, as peripheral regions are generally seen as a residual category, examining differences in entrepreneurial agency between peripheral regions has not been of interest to scholars and thus we do not have sufficiently developed answers to questions regarding these differences. Second, the existing literature on relational aspects in peripheral regions gives some indications that extra-regional networks can compensate possible local disadvantages. Consequently, extra-regional sources of knowledge play a crucial role in successful peripheral development. Third, peripheral regions may possess rich social capital. High levels of social capital can be seen as the basis for an effective collaboration and organization of common regional interests, an aspect that is still barely investigated in entrepreneurship studies dealing with peripheral regions. Not only strong, but also weak ties should be taken into consideration as they could be decisive in explaining differences in competitiveness.

Relational perspectives on entrepreneurship in the periphery

Entrepreneurial heritage

In order to understand and evaluate the heterogeneity of entrepreneurship in peripheral areas, the concept of regional entrepreneurial heritage might prove particularly useful. It assumes that a regional firm population shares a certain set of firm routines.[1] Since external sources of knowledge are crucial for firms in peripheral areas, it is especially important to understand which routines for identifying and acquiring relevant new business knowledge (i.e. absorptive capacity routines, such as firm-internal R&D procedures or strategies for inter-firm communication and cooperation) firms have learned. We further assume that regional entrepreneurial heritage subsequently emerges through inheritance (that is, the transfer of firm routines from a parent firm to its spin-offs) of specific absorptive capacity routines among regional firms, as well as through regional diffusion of these via distinct knowledge exchange channels. Yet, not all local firms will profit in the same way from the diffusion of entrepreneurial heritage. This stands in contrast to basic agglomeration externalities or knowledge spillover concepts, where all firms of a regional economy are supposed to profit from externalities. Rather, differences in absorptive capacity and entrepreneurial agency will be highly accentuated, depending on the firm's history and connectedness. This argumentation is consistent with the heritage theory (Buenstorf & Klepper, 2009), claiming that scale effects and agglomeration externalities are not a prerequisite for the clustering of an industry (Klepper, 2010). Rather, firms with superior routines constitute a particularly fruitful learning environment for entrepreneurs, and spin-offs from these companies are supposed to show a higher chance of survival compared to spin-offs from less successful parent firms or simple start-ups (Dahl & Sorenson, 2013). The fact that spin-offs predominantly locate near their parent firms will ultimately lead to a gradual inheritance and spatial clustering of successful routines within a regional economy. The term regional entrepreneurial heritage stands for this geographically confined accumulation of successful routines.

However, the empirical application of this heritage theory to peripheral areas has been very sparse so far. Only a few studies qualitatively analyzed exceptional spin-off processes from firms located in peripheral areas (Benneworth, 2004; Mayer, 2011). This is surprising because the theoretical assumption of the heritage theory – that agglomeration externalities are not a prerequisite for the clustering of successful firms – makes it well suited for being applied to a peripheral context. There is growing consensus that both organizational inheritance and agglomeration economies give firms a competitive advantage and complementarily contributes to the clustering of an industry. Yet, their relative importance varies depending on the examined industry, its stage in the industry life cycle, and pre-entry experiences of new ventures (Frenken, Cefis & Stam, 2015). This could constitute a specifically interesting field of research, since agglomeration externalities are assumed to be much less significant or even nonexistent in peripheral regions. It would therefore be possible to focus on inheritance mechanisms without a strong influence of agglomeration externalities. However, empirical evidence if and how routine inheritance occurs and how regional entrepreneurial heritage is subsequently built up in peripheral regions is still lacking.

In order to apply the heritage theory to the context of firm population dynamics in the periphery, we need to specify the theory in two aspects: First, we narrow the term routine down to absorptive capacity routines. Absorptive capacity, defined as "a set of organizational routines and processes by which firms acquire, assimilate, transform, and exploit knowledge" (Zahra & George, 2002, p. 186), can be understood as a specific subset of firm routines. The concept is inherently evolutionary, as prior knowledge and capabilities determine the absorptive capacity of a firm, more specifically, its readiness to identify and invest in critical new knowledge (Cohen & Levinthal, 1990). Consequently, the initial conditions under which a firm is founded strongly influences its subsequent development path. Thus, we can assume that the heritage approach, describing routine inheritance more generally, is also relevant for the more specific case of absorptive capacity: Spin-offs would therefore (at least partly) inherit the absorptive capacities of their parent organizations, which would lead to competitive spin-offs from performant parents.

Second, we add routine imitation as another possible process for diffusing absorptive capacity routines within a regional economy. So far, the heritage approach has been focusing on routine inheritance by the creation of new firms, that is, spin-off processes. However, routines, respectively absorptive capacities, are theoretically also diffusible via imitation (Hodgson & Knudsen, 2004). But because of its complexity, the imitation of absorptive capacities is very difficult to perform and is supposed to occur more indirectly through knowledge exchange. From a firm population perspective, the absorptive capacity quality of firms, the similarity of absorptive capacities between exchanging firms and the quality of knowledge network relations determines to what extent firms can learn from each other (Giuliani, 2005). If interactive learning takes place, the enlargement of the knowledge base of the respective firm will consequently have repercussions on its absorptive capacity. Thus, a self-reinforcing, path-dependent process comes into play, where firms with performant absorptive capacities are likely better connected to other firms, which leads to increased learning, which in turn leads to specific modifications of the firms' absorptive capacities (Cohen & Levinthal, 1990). Thus, firms might imitate the absorptive capacities of other firms only to a lesser extent by directly copying it, but depending on what knowledge they acquire through interactive learning. As an example, one could imagine an R&D cooperation between two firms, where the first firm learns of a new promising technology used by the second firm. Without knowing the specificities of this technology, the first firm is now sensitized for the possible benefits of it, and might allocate new investments in order to master it. Thus the first firm modified its absorptive capacity by acquiring new knowledge through inter-firm knowledge exchange. Ultimately, the first firm might add new routines to its existing repertoire, as it masters the new technology. It will then dispose of similar routines, based on the new technology, as the second firm. Hence, routine diffusion is not only possible through routine inheritance, but also through routine imitation based on knowledge exchange.

In the case of firms in peripheral areas, the specified heritage theory might prove particularly useful to analyze their survival performance. As argued before, these firms can rely to a lesser extent on the local knowledge pool and profit less from informal, "random" knowledge spillovers and local agglomeration effects. Thus, they have to rely on their own competencies, on formal relations to extra-regional business partners, and show entrepreneurial agency in order to usefully enlarge their knowledge base. The absorptive capacity then determines how performant firms are in building up these internal competences and firm–external knowledge networks. Evolutionary theory suggests that routines are quite inert to change (Dencker, Gruber & Shah, 2009), and the heritage theory only addresses routine inheritance via spin-offs as mechanism of routine diffusion. Yet, considering the fundamental changes of the economic landscape of the past 20–30 years, globally active firms in peripheral areas must surely have modified their absorptive capacities. It has to be assumed that these firms show entrepreneurial initiative (Cabiddu & Pettinao, 2013) and actively adapt to the changing macroeconomic environment. Considering the lower level of firm formation in peripheral areas, absorptive capacity routines have probably been diffused not only by spin-off processes, but to a certain extent also via imitation. From a firm-centered perspective, the intriguing question is to what extent do firms stick to old routines or apply new ones, and from where they get the routines they inherit or imitate. From a local firm population perspective, the question arises to what extent are different exchange mechanisms in peripheral areas relevant for the diffusion of specific routines, respectively absorptive capacities.

In practice, a few entrepreneurial firms located in peripheral areas may have been able to build up strong absorptive capacities and far-reaching knowledge networks, which make them to a large extent independent from local knowledge exchange. Subsequent investments in knowledge generation and acquisition as well as entrepreneurial discovery will continuously strengthen their absorptive capacities. Nonetheless, these companies might have some business partners in the region with whom knowledge exchange is rational. Trustful and stable relationships might have been built up, as for example in the case of long-term strategic alliances, close buyer–supplier relationships, or interlocking directorates (Carpenter & Westphal, 2001; Westphal, Seidel & Stewart, 2001). Alternatively, senior employees with detailed insights of the company's functioning might move from one firm to the other (Frenken & Boschma, 2007). Also, "technological gatekeepers" (Giuliani, 2005) might play an accentuated role by acquiring external knowledge, facilitating knowledge exchange and building up local capacity in the peripheral context. The dense social relations and high levels of trust among economic actors in peripheral areas might lead to a more pronounced exchange of knowledge und consequently to a more frequent imitation of locally present absorptive capacity routines. With this continuous knowledge exchange, the less competitive firms can make use of the newly acquired knowledge to adapt their absorptive capacities. We can thus speak of the evolution of a specific regional entrepreneurial heritage in peripheral areas. By the processes of inheritance and imitation, specific absorptive capacities are not only a characteristic of a single firm, but to a certain extent also a shared attribute of a regional firm population (Cabiddu & Pettinao, 2013).

Embeddedness

The economically challenging conditions in peripheral regions incentivize firms not only to rely on economic or knowledge networks but also to participate in so-called engagement networks. Successful regional engagement, however, requires entrepreneurial firms to be spatially well embedded.

The embeddedness approach highlights firms' "external relations in specific contexts" (Oinas, 1997, p. 30). Those external relations "may affect the competitiveness of firms and the development of regions" (Oinas, 1997, p. 30). Hence the embeddedness aproach is a promising concept to learn more about the reasons for heterogenous economic development in peripheral regions. Previous studies have shown that embeddedness indeed varies between peripheral areas (Atterton, 2007; Pileček, Chromý & Jančák, 2013). When examining peripheral regions, it is crucial to take into consideration not only the embeddedness of firms at a regional level, but also at an extra-regional level, which has often been neglected in embeddedness research (Hess, 2004). Spatial embeddedness (see Oinas, 1997) however highlights the entrepreneurs' social networks covering different spatial levels. This is especially important concerning regional engagement in peripheral regions, as a good mix of regional networks to other actors such as other entrepreneurs or the municipal authorities and extra-regional networks to decision-makers who are often located in core regions, may be advantageous to firms in the periphery.

Granovetter (1990, p. 98) underlines the potential of actors to shape their environment by defining embeddedness as follows: "By 'Embeddedness' I mean that economic action, outcomes, and institutions are affected by actor's personal [dyadic] relations, and by the structure of the overall network of relations." That means that economic actors are embedded in specific contexts (see Welter, 2011) such as organizational or institutional contexts, which in turn can be shaped through engagement networks. Examples of such networks are networks formed for collective action within regional business associations or the networks resulting from the entrepreneur's membership in a political party, but also informal networks between entrepreneurs who meet to engage for common regional interests.

So far, many studies on embeddedness in entrepreneurship have focused on knowledge and information networks mostly concerning business or innovation issues (Johannisson, Ramirez-Pasillas & Karlsson, 2002; Uzzi, 1996), not however on networks for regional engagement. Regional engagement, in the sense of entrepreneurs who actively shape the contexts they are involved in (see Lengauer & Tödtling, 2010), depends on those networks at different spatial scales (spatial embeddedness). Moreover, the characteristics and the effectiveness of engagement networks differ between peripheral regions. In this sense entrepreneurial actors in peripheral regions can actively shape and modify the contexts they are embedded in. Welter (2011) distinguishes between business (industry), spatial (business support infrastructure), social (networks) and institutional (legal and regulatory regulations) contexts. In peripheral regions certain contexts such as the institutional, but also the organizational contexts are of particular interest, as by shaping them, entrepreneurs may possibly reduce the organizational and institutional thinness[2] of their region.

As a result, we note that not only knowledge networks for business purposes in the strict sense are worth examining, but also networks with the aim of shaping regional contexts. By taking such a perspective, studies of entrepreneurship in peripheral regions may cover a broader spectrum of relations, which in turn may have explanatory capacity concering differences in terms of entrepreneurial dynamics between peripheral regions.

Lengauer and Tödtling (2010, p. 2) examine regional engagement in the sense of corporate regional engagement, which they define as "the active involvement of companies in shaping and upgrading regional productive potentials." It means that entrepreneurs actively shape "the contexts and networks a firm is involved in" (Lengauer & Tödtling, 2010, p. 7). In their study of corporate regional engagement, Lengauer and Tödtling (2010) do not focus on the networks this engagement is based on, but rather on the motivation for and the degree of corporate regional engagement by comparing three different industries in the Austrian region of Styria. Moreover they focus on different activities the regions benefit from, such as human resource development and training or philanthropic activities (Lengauer & Tödtling, 2010). Yet, they do not explicitly take into consideration the engagement of firms to shape institutional issues such as laws or regulations, which also influence regional economies. Hence it makes sense to consider different contexts entrepreneurs shape through regional engagement, such as organizational, social or spatial contexts for example, but also institutional contexts (see Welter, 2011). Regarding the conceptual approach we develop in this chapter, regional engagement includes all kinds of activities and networks shaping regional contexts, which have direct or indirect economic effects and are therefore eventually business oriented.

When engaging for the region, networks of different scales and actors are indespensible for entrepreneurs. To shape institutional contexts extra-regional relations are particularly crucial as decision-makers are often located in the core. To shape organizational contexts, the willingness of firms to collaborate regionally is of utmost importance. Hence collective action among several entrepreneurs, but also between entrepreneurs and actors of other interest groups (such as the municipal authorities or tourism) at a regional level, is significant and seems to have an important influence on regional development (Engstrand & Sätre Åhlander, 2008).

Focusing on regional engagement of entrepreneurs in peripheral regions and their spatial embeddedness is important for several reasons: First, according to Baumgartner, Pütz and Seidl (2013) one typical characteristic of entrepreneurship in European peripheral regions is that it "aims to create added values locally" (Baumgartner, Pütz & Seidl, 2013, p. 18). It means that entrepreneurs in peripheral areas show a strong willingness to engage on behalf of the regional economy (Baumgartner, Pütz & Seidl, 2013). This engagement may however differ between peripheral regions. Such differences in regional engagement may help explain different development dynamics. Second, regional engagement benefits from the rich social capital[3] that can be found in peripheral regions. As mentioned earlier, actors in peripheral regions possess strong, especially informal, relations that are based on high levels of trust (Atterton, 2007). That means that entrepreneurial actors in peripheral regions know each other and it is easy for them to have face-to-face contact. Due to the small population size that is typical for peripheral regions, actors can meet quickly, often without passing through official channels, and they may develop a capacity to react to issues in flexible and effective ways. Although dense social networks can help entrepreneurial actors in peripheral regions to act quickly and to collaborate efficiently to foster regional economies, it is essential that they not only develop strong ties (bonding social capital) but also weak ties (bridging social capital) to other regional actors. To engage regionally, different regional actors have to be willing to collaborate. The quality of collaboration is very important, especially for entrepreneurship in peripheral regions (Pato & Teixeira, 2014). If there is a lack of reciprocity or collaboration between firms and other regional stakeholders, they risk to negatively influence long-term regional economies. Moreover, entrepreneurial actors should dispose of extra-regional ties (linking social capital), as decision-makers are often located in core regions. This is especially important when entrepreneurs try to shape institutional contexts such as laws or regulations. Hence regional and extra-regional networks are important at the same time. If there is a lack of extra-regional or weak ties or the willingness to collaborate, peripheral regions may be in danger of lock-in situations. Therefore, both a combination of strong ties (bonding social capital) and weak ties (bridging social capital) connecting the firms with regional and extra-regional networks (linking social capital) are important. In addition, collaboration between and among different regional actors is another prerequisite for regional engagement to be successful.

Engagement networks very often include the participation not only of several entrepreneurial actors, but also of several agents from different sectors of society. Hence the concept of cross-sectoral social capital, which highlights the importance of a collective of actors (Westlund & Gawell, 2012), is suitable to examine those networks. Engagement networks aimed at shaping regional contexts rest upon the participation of several actors, thus bonding, bridging and linking social capital is needed (Westlund & Gawell, 2012) and these different types of social capital become effective at different spatial scales. As already mentioned, bonding social capital is very important in the context of peripheral regions. To engage for common regional interests, bonding social capital at the regional level, e.g. in the form of business organizations or informal networks between entrepreneurs, is important. But also the so-called weak ties or the bridging social capital at the regional level are indispensable, consisting of relations to other important regional stakeholders such as the public authorities, especially when engaging for the region. Linking social capital however involves actors at a higher administrative hierarchical level, who are crucial when shaping, for example, institutional contexts such as laws or regulations. Since these administrative hierarchical levels are almost exclusively located in core regions, linking social capital of peripheral entrepreneurs is practically always related to extra-regional ties.

Therefore highly competitive peripheral regions may show well developed regional engagement of entrepreneurs based on high social capital and the ability of the entrepreneurs to organize and collaborate at a regional level, involving bonding and bridging social capital. Additionally, linking social capital at the extra-regional scale is well developed. In contrast, economically less competitive peripheral regions may be characterized by too many strong and local links or a weakly developed willingness to collaborate, i.e. they face the risk of so-called over-embeddedness (Uzzi, 1996) which may result in lock-in (Atterton, 2007). That also means that they do not have enough extra-regional linking social capital connecting them to important decision-makers at different hierarchical levels when institutional contexts are to be shaped. Thanks to regional engagement, entrepreneurial actors can influence, for example, aspects of organizational (e.g. universities, associations) and institutional thinness (e.g. laws, rules, cooperation culture) to a certain degree. Following such a perspective may explain why peripheral regions differ.

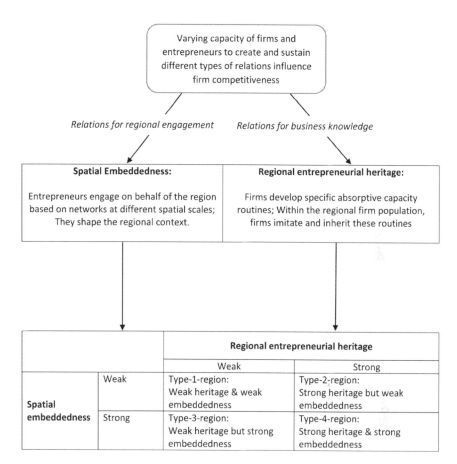

Figure 9.1 Conceptual framework for a relational perspective on entrepreneurship in the periphery

Conceptual framework and possible policy implications

Based on the preceeding conceptual discussion, we propose to examine the two presented approaches in a combined way, as both types are potentially relevant when it comes to the competitiveness of firms in peripheral areas. More specifically, the competitiveness depends on their capacity to construct and sustain both relations for regional engagement and for business knowledge acquisition. As we outline above, we suggest to employ the heritage approach to investigate business knowledge networks and the embeddedness perspective to examine engagement networks. By differentiating the quality of these knowledge and engagement networks, we deduct a typology of four types of peripheral regional economies (see Figure 9.1). We further specify these types of peripheral economies by describing several relevant aspects related to the embeddedness and heritage perspectives and compare them to an archetypical core economy (see Table 9.1).

Table 9.1 Key heritage and embeddedness characteristics of the different types of peripheral regions

Region-type/ characteristics	Type-1 region: weak heritage and weak embeddedness	Type-2 region: strong heritage but weak embeddedness	Type-3 region: weak heritage but strong embeddedness	Type-4 region: strong heritage and strong embeddedness	Core regions
Regional routine inheritance dynamics	Start-up activities from experienced entrepreneurs and spin-off dynamics are rare	Average start-up activities from experienced entrepreneurs and spin-off dynamics	Start-up activities from experienced entrepreneurs and spin-off dynamics are rare	Average start-up activities from experienced entrepreneurs and spin-off dynamics	High and self-reinforcing spinoff dynamics lead to a stronger diffusion of successful routines
Regional routine imitation dynamics	Few economic links between regional firms lead to very limited knowledge exchange and routine imitation	Firms with strong routines collaborate with selected regional firms; limited routine imitation dynamics	Few economic links between regional firms lead to very limited knowledge exchange and routine imitation	Firms with strong routines collaborate with selected regional firms; limited routine imitation dynamics	Many firms with strong routines and intensive knowledge exchange lead to high routine imitation dynamics
Absorptive capacity routines and knowledge networks	A small minority of firms dispose of successful routines and strong knowledge networks; these are hardly dependent on the regional firm population	Some firms dispose of successful routines and knowledge networks; a limited number of firms share a strong regional entrepreneurial heritage	A small minority of firms dispose of successful routines and strong knowledge networks; these are hardly dependent on the regional firm population	Some firms dispose of successful routines and knowledge networks; a limited number of firms share a strong regional entrepreneurial heritage	High competition and strong knowledge exchange lead to the diffusion and selection of the fittest routines

Value creation for the region	Entrepreneurial actors only engage weakly for the region	Entrepreneurial actors only engage weakly for the region	Entrepreneurial actors engage strongly for the region	Entrepreneurial actors engage strongly for the region	Entrepreneurial actors do not especially engage for the region, as the contexts in core regions are already favorable
Bonding and bridging social capital and degree of collaboration	Too much bonding social capital of entrepreneurial actors involves the risk of lock-in; they dispose of only weak bridging social capital and a weak willingness to collaborate	Too much bonding social capital of entrepreneurial actors involves the risk of lock-in; they dispose of only weak bridging social capital and a weak willingness to collaborate	Entrepreneurial actors dispose of a good mix of bonding and bridging social capital and a strong willingness to collaborate	Entrepreneurial actors dispose of a good mix of bonding and bridging social capital and a strong willingness to collaborate	Entrepreneurial actors dispose of high bonding and bridging social capital, but not focused on regional engagement, but on business collaboration
Linking social capital	Entrepreneurial actors lack extra-regional relations (linking social capital) to decision-makers in the core	Entrepreneurial actors lack extra-regional relations (linking social capital) to decision-makers in the core	Entrepreneurial actors dispose of a good mix of regional and extra-regional (linking social capital) relations	Entrepreneurial actors dispose of a good mix of regional and extra-regional (linking social capital) relations	Entrepreneurial actors dispose of higher linking social capital and this at all spatial scales

Type-1 regions are those regional economies that many scholars would identify as typically peripheral. Indeed, they face the highest challenges, as entrepreneurial heritage and regional engagement are weakly developed. Regional firm populations might be characterized by traditional industrial sectors with low export and innovation rates. Moreover, due to limited regional engagement, organizational and institutional thinness persist.

Type-2 regions are characterized by a reasonable amount of competitive and well connected firms. Some firms benefit from regional routine inheritance or imitation. However, these firms may be only loosely anchored in the region and may not see the necessity for regional engagement. This might be the case for newly immigrated entrepreneurs or foreign-owned firms.

Type-3 regions are characterized by weaknesses concerning entrepreneurial heritage characteristics. Regional routine inheritance and imitation dynamics as well as knowledge networks and absorptive capacities of firms are restricted. This may be due to a firm population that is mainly active in mature industries and disposes of weakly performant knowledge networks when it comes to creating innovations. On the other hand, regional engagement is more pronounced. The entrepreneurs can rely on a good mixture of bonding, bridging and linking social capital, including extra-regional relations which may be due to a long industrial tradition and a well-established organizational framework in the region.

Type-4 regions host the most successful peripheral economies as both knowledge and engagement networks are well developed and efficient. The firms in these regions generally possess a high capacity to create and sustain those relevant networks, which help them compensate the economically challenging conditions in peripheral regions. However, they differ from core regions in two important respects: First, the limited size of the firm population hinders the take-off of a self-reinforcing clustering process. In this sense, firms in peripheral regions will still have to rely to a larger degree on external knowledge sources compared to firms in core regions. Second, as core regions are generally better positioned when it comes to shaping their regional economic contexts, firms and entrepreneurs from peripheral regions still have to advocate for better conditions on behalf of their regions.

In order to address these specific deficiencies of different peripheral regional economies, differentiated policy approaches are necessary. As Type-1 regions lack efficient knowledge and engagement networks, regional policy should primarily aim at identifying those actors that have the highest potential for strengthening knowledge networks – both regional and extra-regional – and building up regional engagement alliances. Since Type-2 regions show deficiencies concerning the firms' regional engagement, the well connected and competitive firms should be encouraged to use their network competencies on behalf of the regional contexts. This should be done by integrating other firms and regional actors, and by creating engagement networks at an extra-regional scale. Type-3 regions are marked by deficiencies concerning entrepreneurial heritage. Hence, firms' social capital for regional engagement could serve as a basis for fostering knowledge exchange at both regional and extra-regional levels. Finally, Type-4 regions are the most successful ones as both knowledge networks and engagement networks are well developed. Therefore policy action should strengthen already well established and further promising business fields in order to surpass the critical mass for self-reinforcing cluster dynamics.

Conclusions

This chapter showed that entrepreneurship in peripheral areas should be analyzed in more depth in order to identify and understand the heterogeneity of regional economic development in the periphery. It is important to note that a modern definition of peripherality has to go beyond simple geographical distance, but has to take into account other forms of proximity, especially when it comes to the connectedness of regional economies, both at the local and the global level. As current entrepreneurship approaches cannot explain these heterogeneities between peripheral regions adequately, we have to examine peripheral economies and their differences from another angle, especially when taking into account that firms in the periphery are also embedded in global production and distribution networks while at the same time being rooted in a peripheral context. Hence a relational firm-centered perspective should be employed, as it takes into account the situation that peripheral areas are confronted with today. Based on relevant findings from previous literature we come to the following conclusions: As we argue above, entrepreneurial firms in peripheral regions can compensate lacking scale effects and institutional and organizational thinness. This can be done by fostering extra-regional relations and by engaging for the region combining bonding, bridging and linking ties. It is thus necessary to better understand the network strategies of these firms and to explore how the knowledge and engagement networks of these firms evolve over time. Employing a heritage perspective enables us to examine the dynamics of firm routines and absorptive capacity diffusion. Additionally, employing an embeddedness perspective has the advantage to investigate a different purpose of network relations with the aim of shaping regional contexts in the periphery.

We suggest combining the analysis of both knowledge and engagement networks as we assume that the competencies necessary to engage in these networks are similar, and that both influence the competitiveness of firms. The combination of those two perspectives has therefore great explanatory potential concerning the economic heterogeneity of peripheral regions. Based on different qualities of knowledge and engagement networks, we developed a typology of four peripheral regional economies and deducted some basic policy recommendations. As policies tend to suggest "one-size-fits-all" solutions (Tödtling & Trippl, 2005) for peripheral regions without taking into consideration their differences, more sophisticated and differentiated approaches to support the development efforts of these regions would be welcomed by theorists and practitioners alike.

With this combined approach we also intend to contribute to the discussions in relational economic geography, which "draws attention to the importance of economic agents and how they act and interact in space" (Bathelt & Glückler, 2003, p. 128). Firms can shape the regional economic context by building different types of networks. The context is thus not a preformed and unchangeable attribute of the regional economy, but an emergent property resulting from the specific regional actor constellation. Consequently, it is not primarily the region's characteristic as "peripheral" that determines entrepreneurial agency, but the different ways entrepreneurial actors build relations and employ them to shape their environment. Since we see the context as an emergent property resulting from network dynamics, it is essential to take into consideration an evolutionary perspective. This enables the integration of distinct explanatory evolutionary concepts, such as path-dependence or contingency (Bathelt & Glückler, 2003). In this chapter, we attempted to construct a conceptual framework that gives more attention to economic actors in peripheral regions, the way they integrate in different types of regional and extra-regional networks and how this might influence regional economic development.

Future empirical studies are necessary to verify this conceptual framework. Since the framework is based on individual firms and entrepreneurs, detailed micro-scale data, which is not always available, is necessary. In order to get a good picture of both knowledge and engagement networks, there is a need to combine different data. These may include R&D cooperation, joint patent pending, interlocking directorates and spin-off–parent relations when speaking of knowledge networks, or memberships in business organizations and political parties as well as political mandates held by entrepreneurs in the context of engagement networks, to name just a few. Besides the specific characteristics and functioning principles of both types of networks in different peripheral regions, analyzing the interplay and co-evolution of knowledge and engagement networks represents another intriguing avenue of research. Questions such as whether both types of networks are co-evolving or whether they are independent from each other are relevant. If they are indeed co-evolving, is one type of network more dependent on the other, i.e. is the evolution of one type of network a prerequisite for the development of the other type? Of course, many other interesting research questions are possible within this research field, which might give valuable insights on the different development patterns of peripheral regions.

Acknowledgement

The research for this chapter was funded by the Swiss National Science Foundation (Grant 146436). The authors would like to thank the editors for their helpful comments and feedback.

Notes

1 A firm routine can be defined as "an executable *capability* for repeated performance in some *context* that [has] been *learned* by an organization in response to *selective pressures*" (Cohen et al., 1996, p. 683).
2 Trippl, Asheim and Miörner (2015, p. 5) define organizational thickness (thinness) as "the presence (absence) of a critical mass of firms, universities, research bodies, support organizations, unions, associations, and so on." Institutional thickness (thinness) refers to "the presence (absence) of both formal institutions (laws, rules, regulations) and informal institutions such as innovation and cooperation culture, norms and values that promote collective learning and knowledge exchange" (Trippl, Asheim & Miörner, 2015, p. 5).
3 Social capital can be defined as consisting of "social networks/relations and the norms and values that are generated, accumulated and disseminated through these networks" (Westlund & Gawell, 2012, p. 104).

References

Anderson, A. R. (2000). Paradox in the periphery: an entrepreneurial reconstruction? *Entrepreneurship & Regional Development*, 12(2), 91–109.

Atterton, J. (2007). The "strength of weak ties": social networking by business owners in the Highlands and Islands of Scotland. *Sociologia Ruralis*, 47(3), 228–245.

Audretsch, D. B., Falck, O., Feldman, M. P. & Heblich, S. (2012). Local entrepreneurship in context. *Regional Studies*, 46(3), 379–389.

Bathelt, H. & Glückler, J. (2003). Toward a relational economic geography. *Journal of Economic Geography*, 3(2), 117–144.

Bathelt, H., Malmberg, A. & Maskell, P. (2004). Clusters and knowledge: local buzz, global pipelines and the process of knowledge creation. *Progress in Human Geography*, 28(1), 31–56.

Baumgartner, D., Pütz, M. & Seidl, I. (2013). What kind of entrepreneurship drives regional development in European non-core regions? A literature review on empirical entrepreneurship research. *European Planning Studies*, 21(8), 1095–1127.

Benneworth, P. (2004). In what sense "regional development?": entrepreneurship, underdevelopment and strong tradition in the periphery. *Entrepreneurship & Regional Development*, 16(6), 439–458.

Boschma, R. & Lambooy, J. G. (1999). Evolutionary economics and economic geography. *Journal of Evolutionary Economics*, 9(4), 411–429.

Bosma, N. & Schutjens, V. (2011). Understanding regional variation in entrepreneurial activity and entrepreneurial attitude in Europe. *The Annals of Regional Science*, 47(3), 711–742.

Buenstorf, G. & Klepper, S. (2009). Heritage and agglomeration: the Akron tyre cluster revisited. *The Economic Journal*, 119(April), 705–733.

Cabiddu, F. & Pettinao, D. (2013). External knowledge, territorial inertia and local development: an explanatory case study. *European Planning Studies*, 21(8), 1297–1316.

Carpenter, M. A. & Westphal, J. D. (2001). The strategic context of external network ties: examining the impact of director appointments on board involvment in strategic decision making. *The Academy of Management Journal*, 44(4), 639–660.

Cohen, M. D., Burkhard, R., Dosi, G., Egidi, M., Marengo, L., Warglien, M. & Winter, S. (1996). Routines and other recurring action patterns of organizations: contemporary research issues. *Industrial and Corporate Change*, 5(3), 653–698.

Cohen, W. M. & Levinthal, D. A. (1990). Absorptive capacity: a new perspective on learning and innovation. *Administrative Science Quarterly*, 35(1), 128–152.

Copus, A., Skuras, D. & Tsegenidi, K. (2008). Innovation and peripherality: an empirical comparative study of SMEs in six European Union countries. *Economic Geography*, 84(1), 51–82.

Dahl, M. S. & Sorenson, O. (2013). The who, why, and how of spinoffs. *Industrial and Corporate Change*, 23(3), 661–688.

Dencker, J. C., Gruber, M. & Shah, S. K. (2009). Pre-entry knowledge, learning, and the survival of new firms. *Organization Science*, 20(3), 516–537.

Doloreux, D. (2003). Regional innovation systems in the periphery: the case of the Beauce in Québec (Canada). *International Journal of Innovation Management*, 7(1), 67–94.

Engstrand, Å. K. & Sätre Åhlander, A. M. (2008). Collaboration for local economic development: business networks, politics and universities in two Swedish cities. *European Planning Studies*, 16(4), 487–505.

Freire-Gibb, L. C. & Nielsen, K. (2014). Entrepreneurship within urban and rural areas: creative people and social networks. *Regional Studies*, 48(1), 139–153.

Frenken, K. & Boschma, R. (2007). A theoretical framework for evolutionary economic geography: industrial dynamics and urban growth as a branching process. *Journal of Economic Geography*, 7(5), 635–649.

Frenken, K., Cefis, E. & Stam, E. (2015). Industrial dynamics and clusters: a survey. *Regional Studies*, 49(1), 10–27.

Giuliani, E. (2005). Cluster absorptive capacity: why do some clusters forge ahead and others lag behind? *European Urban and Regional Studies*, 12(3), 269–288.

Grabher, G. (1993). The weakness of strong ties: the lock-in of regional development in the Ruhr area. In G. Grabher (Ed.), *The Embedded Firm* (pp. 255–277). London: Routledge.

Granovetter, M. (1973). The strength of weak ties. *American Journal of Sociology*, 78(6), 1360–1380.

Granovetter, M. (1990). The old and new economic sociology: a history and an agenda. In R. Friedland & A. Robertson (Eds.), *Beyond the Market Place: Rethinking Economy and Society* (pp. 89–112). New York: Aldine de Gruyter.

Grillitsch, M. & Nilsson, M. (2015). Innovation in pheripheral regions: do collaboratins compensate for a lack of local knowledge spillovers? *The Annals of Regional Science*, 54, 299–321.

Hess, M. (2004). "Spatial" relationships? Towards a reconceptualization of embeddedness. *Progress in Human Geography*, 28(2), 165–186.

Hodgson, G. M. & Knudsen, T. (2004). The firm as an interactor: firms as vehicles for habits and routines. *Journal of Evolutionary Economics*, 14(3), 281–307.

Isaksen, A. (2015). Industrial development in thin regions: trapped in path extension? *Journal of Economic Geography*, 15(3), 585–600.

Johannisson, B., Ramirez-Pasillas, M. & Karlsson, G. (2002). The institutional embeddedness of local inter-firm networks: a leverage for business creation. *Entrepreneurship & Regional Development*, 14(4), 297–315.

Klepper, S. (2010). The origin and growth of industry clusters: the making of Silicon Valley and Detroit. *Journal of Urban Economics*, 67(1), 15–32.

Lagendijk, A. & Lorentzen, A. (2007). Proximity, knowledge and innovation in peripheral regions: on the intersection between geographical and organizational proximity. *European Planning Studies*, 15(4), 457–466.

Lengauer, L. & Tödtling, F. (2010). Regional embeddedness and corporate regional engagement: evidence from three industries in the Austrian region of Styria. In *Conference paper for the 8th European Urban & Regional Studies Conference* (pp. 1–31).

Maillat, D., Lecoq, B., Nemeti, F. & Pfister, M. (1995). Technology district and innovation: the case of the Swiss Jura Arc. *Regional Studies*, 29(3), 251–263.

Malecki, E. (1994). Entrepreneurship in regional and local development. *International Regional Science Review*, 16(1–2), 119–153.

Mayer, H. (2011). *Entrepreneurship and Innovation in Second Tier Regions.* Cheltenham: Edward Elgar.

Meccheri, N. & Pelloni, G. (2006). Rural entrepreneurs and institutional assistance: an empirical study from mountainous Italy. *Entrepreneurship & Regional Development*, 18(5), 371–392.

Metcalfe, S. (2004). The entrepreneur and the style of modern economics. *Journal of Evolutionary Economics*, 14(2), 157–175.

OECD (2006). *The New Rural Paradigm: Policies and Governance.* Paris: OECD Publishing.

Oinas, P. (1997). On the socio-spatial embeddedness of business firms. *Erdkunde*, 51(1), 23–32.

Pato, M. L. & Teixeira, A. (2014). Twenty years of rural entrepreneurship: a bibliometric survey. *Sociologia Ruralis*, 1–26.

Pileček, J., Chromý, P. & Jančák, V. (2013). Social capital and local socio-economic development: the case of Czech peripheries. *Tijdschrift Voor Economische En Sociale Geografie*, 104(5), 604–620.

Putnam, R. D. (2000). *Bowling Alone: The Collapse and Revival of American Community.* New York: Simon & Schuster.

Rodríguez-Pose, A. & Fitjar, R. D. (2013). Buzz, archipelago economies and the future of intermediate and peripheral areas in a spiky world. *European Planning Studies*, 21(3), 355–372.

Schutjens, V. & Stam, E. (2003). The evolution and nature of young firm networks: a longitudinal perspective. *Small Business Economics*, 21(2), 115–134.

Shane, S. (2003). *A General Theory of Entrepreneurship: The Individual-Opportunity Nexus.* Cheltenham: Edward Elgar.

Simon, H. (2009). *Hidden Champions of the Twenty-First Century: The Success Strategies of Unknown World Market Leaders.* New York: Springer.

Tödtling, F. & Trippl, M. (2005). One size fits all? Towards a differentiated regional innovation policy approach. *Research Policy*, 34(8), 1203–1219.

Torre, A. & Gilly, J.-P. (2000). On the analytical dimension of proximity dynamics. *Regional Studies*, 34(2), 169–180.

Trippl, M., Asheim, B. & Miörner, J. (2015). Identification of regions with less developed research and innovation systems research and innovation systems. *CIRCLE Papers in Innovation Studies*, 1, 1–22.

Uzzi, B. (1996). The sources and consequences of embeddedness for the economic performance of organizations: the network effect. *American Sociological Review*, 61(4), 674–698.

Vaillant, Y. & Lafuente, E. (2007). Do different institutional frameworks condition the influence of local fear of failure and entrepreneurial examples over entrepreneurial activity? *Entrepreneurship & Regional Development*, 19(4), 313–337.

Virkkala, S. (2007). Innovation and networking in peripheral areas – a case study of emergence and change in rural manufacturing. *European Planning Studies*, 15(4), 511–529.

Ward, N. & Brown, D. L. (2009). Placing the rural in regional development. *Regional Studies*, 43(10), 1237–1244.

Welter, F. (2011). Contextualizing entrepreneurship – conceptual challenges and ways forward. *Entrepreneurship Theory and Practice*, 35(1), 165–184.

Westlund, H. & Bolton, R. (2003). Local social capital and entrepreneurship. *Small Business Economics*, 21(2), 77–113.

Westlund, H. & Gawell, M. (2012). Building social capital for social entrepreneurship. *Annals of Public and Cooperative Economics*, 83(1), 101–116.

Westphal, J. D., Seidel, M.-D. L. & Stewart, K. J. (2001). Second-order imitation: uncovering latent effects of board network ties. *Administrative Science Quarterly*, 46(4), 717–747.

Zahra, S. & George, G. (2002). Absorptive capacity: a review, reconceptualization, and extension. *Academy of Management Review*, 27(2), 185–203.

10 The geography of entrepreneurship

Where are we? Where do we go?

Haifeng Qian

Introduction

The focus of this book is on entrepreneurship from a geographical perspective. Within this scope, the chapters included have exhibited high diversities in terms of entrepreneurship measures (i.e., entrepreneurs, self-employment, traditional start-ups, and technology ventures), countries or regions (i.e., Canada, China, EU, Germany, Korea, Sweden, the United States), theoretical perspectives, and methodologies. The book represents a timely collection that continues and advances scholarly understanding on the geography of entrepreneurship. In this concluding chapter, I recapitulate the contributions of these chapters in a broad context of the literature, and propose three directions for future research on the geography of entrepreneurship.

Recapitulation: contributions of this book in a broad context of geographical studies of entrepreneurship

The past two or three decades have witnessed the emergence and flourishing of the literature on the geography of entrepreneurship. While largely interdisciplinary, the topic is particularly popular in the fields of regional science and regional studies. For instance, the four decennial special issues on entrepreneurship in the journal *Regional Studies* are well received. The large body of literature along this line of research can be primarily divided into two sub-topics: (1) the role of entrepreneurship in regional economic development and (2) geographical factors leading to regional variations in entrepreneurship. The first sub-topic helps to understand why it is important to study the geography of entrepreneurship. Accepting the importance of entrepreneurship, the second sub-topic sheds light on how to improve regional environmental factors so as to build an entrepreneurial regional economy.

On the first sub-topic, the role of entrepreneurship in regional economic development is well recognized (Malecki, 1993), though depending on the measures of both entrepreneurship and economic development. For instance, new firm formation is one of the key drivers of regional employment growth (Acs & Armington, 2006) and accounts for almost all the net job growth in the U.S. context (Haltiwanger et al., 2013). Besides employment growth, the productivity growth effect of entrepreneurship has also drawn much attention, usually in the context of innovation and knowledge- or technology-driven economies. Although Schumpeter (1934) addressed the role of entrepreneurs in product and process innovations over eight decades ago, it was not until the recent research on the geography of entrepreneurship that scholars and practitioners recognized the contribution of entrepreneurship to regional productivity via innovation. Technology entrepreneurs often create new firms to commercialize knowledge created in research or private organizations in a region (Audretsch & Lehmann, 2005; Qian et al., 2013). By doing that, they introduce innovations into the market and further improve regional productivity. Not surprisingly, many new technology firms are spin-offs of incumbents, which, according to Klepper (2010), gave rise to the industrial clusters in Detroit and Silicon Valley. Because of both job creation and productivity growth effects, policy makers across countries and regions have shown great interests in supporting technology ventures, with the hope of creating the next Silicon Valley. Without exception, this is also the case in South Korea. In Chapter 5, Koo and Yoo examine the entrepreneurship policies and budgets both at central government level and for the most entrepreneurial regions in Korea (including the national capital, Seoul). They find that entrepreneurship policies are designed overwhelmingly for technology ventures in contrast to traditional small businesses. Their data, on the other side, reveal that technology ventures account for a very small share of start-ups (2 percent nationwide and 4 percent in Seoul). With this mismatch, Koo and Yoo suggest policy makers should "pay more attention to small business."

On the second sub-topic, regional variations in entrepreneurship have also been extensively studied, identifying a number of geographically bounded factors associated with entrepreneurship. For instance, Reynolds et al. (1994) note the impact of the following factors on new firm formation: demand (e.g., from population growth), agglomeration, unemployment, wealth, small business concentration, political ethos and public policy. These factors have been widely tested in empirical studies since then. It is worth noting that unemployment more likely is conducive to less productive "necessity entrepreneurship" (Acs, 2006), i.e., the fact that people are forced to start their businesses because they cannot get other jobs. In addition, entrepreneurial activity appears to be path dependent; historically more entrepreneurial regions are more likely to remain entrepreneurial (Andersson & Koster, 2011; Fritsch & Wyrwich, 2014). Recently, the focus on knowledge- or technology-based entrepreneurship has shifted the discussion to a new set of regional factors, including but not limited to new knowledge or organizations generating new knowledge (Acs et al., 2009; Audretsch et al., 2013; Smith & Bagchi-Sen, 2012), human capital (Acs & Armington, 2004; Qian et al., 2013), diversity (Audretsch et al., 2010; Lee et al., 2004; Qian, 2013), soft infrastructure such as incubators and broadband (Aernoudt, 2004; Mack, 2014), institutions (Acs et al., 2008; Welter, 2011), public policy (Minniti, 2008; Qian & Haynes, 2014), and networking/social capital (Feldman & Zoller, 2012; Nijkamp, 2003; Westlund & Bolton, 2003), among others.

Most of the chapters in this book are in line with and advance this second sub-topic on regional variations in entrepreneurship. In Chapter 2, Capello and Lenzi demonstrate that regional entrepreneurship depends on "territorial patterns of innovation" (Capello, 2013). It has been well documented in the literature (e.g., Acs et al., 2009) that new knowledge (especially technologically oriented) is one source of entrepreneurial opportunities. However, the innovative environment under which entrepreneurs commercialize new knowledge is not well understood. The work by Capello and Lenzi provides one perspective towards this direction. Both Chapter 3 by Spigel and Chapter 9 by Bürcher, Habersetzer, and Mayer focus its topic on the role of networking in entrepreneurial behavior in different geographical contexts. Each advances our understanding on different types of entrepreneurial networks that are geographically embedded. In Chapter 4, Fritsch and Wyrwich move further their previous research (Fritsch & Wyrwich, 2014) on the path-dependency nature of entrepreneurship. They report how history matters differently for different types of self-employment. He, Guo, and Zhu in Chapter 6 reveal how institutional changes have unleashed entrepreneurship in China. They also demonstrate the divergent entrepreneurial performance between Chinese cities that have experienced greater institutional reforms in terms of marketization, decentralization, and globalization and those have experienced lower. Chapter 7 by Mack and Credit and Chapter 8 by Andersson, Koster, and Lavesson are both related to the relationship between agglomeration and entrepreneurship. Mack and Kevin document spatial-temporal dynamics of entrepreneurship of ten U.S. metropolitan areas. Their focus on intra-metropolitan distributions of entrepreneurship makes the chapter one of the few studies that offer insights into this important urban development issue. Using Swedish data, Andersson et al. provide additional evidence that knowledge-based entrepreneurship is more concentrated in urban areas than rural areas, benefiting from agglomeration economies of the former.

Moving forward: three promising research directions on the geography of entrepreneurship

The geography of entrepreneurship is still a fast-growing research area. Recent developments, including some theoretical or empirical studies from the chapters in this book, have revealed a few promising directions that not only contribute to better understandings of the geographical nature of entrepreneurship, but also have important implications on regional economic development policy. These deserve much additional work. Below I discuss three directions for future studies.

The geography of knowledge spillover entrepreneurship

The importance of knowledge to entrepreneurship, as discussed by Capello and Lenzi in Chapter 2, is well recognized. One of the core theories along this line of research is the knowledge spillover theory of entrepreneurship (Acs et al., 2009; Audretsch & Lehmann, 2005). The theory points to the role of entrepreneurs in commercializing knowledge that is locally created in universities or incumbent firms but left commercially unexploited. They start new businesses to explore these potential market opportunities. This is consistent with the spin-off literature by Klepper (2010). While the theory is straightforward, the entrepreneurial process of knowledge commercialization has not been well understood. Questions that remain to be answered include:

1. How can regional knowledge bases be measured? What types of knowledge bases are more conducive to entrepreneurship?
2. What types of skills at the regional level are needed to facilitate the knowledge commercialization process by entrepreneurs?
3. What types of social, economic, and institutional environments at the regional level can benefit knowledge spillover entrepreneurship?

Research on these three questions is emerging, but far from sufficient. For question 1, it is indeed challenging to categorize and measure regional knowledge bases. Asheim and his collaborators (Asheim et al., 2011; Asheim & Coenen, 2005; Asheim et al., 2007) have made one notable effort toward this direction by building a typology that includes three knowledge bases: analytical (science-based), synthetic (engineering-based), and symbolic (arts-based). However, the relationships between these different knowledge bases and entrepreneurship have not been studied. For question 2, Qian and Acs (2013) argue that knowledge itself (as entrepreneurial opportunities) will not lead to a vibrant entrepreneurial regional economy without sufficient *entrepreneurial absorptive capacity*, i.e., entrepreneurs' skills needed to commercialize new knowledge. Yet such skills have not been well identified and tested at the regional level. For question 3, the work of Capello and Lenzi in this book is a good start by looking at the impact of territorial patterns of innovations on entrepreneurship. I will argue that more comprehensive efforts are needed. It is necessary to study the *"entrepreneurial milieu"* under which local entrepreneurs with sufficient absorptive capacity can smoothly exploit the market value of new knowledge without social, economic, and institutional barriers.

The three questions above can be combined into a holistic approach to knowledge spillover entrepreneurship, as demonstrated in Figure 10.1. This combination helps build a through understanding of knowledge-driven entrepreneurship, and further help develop a comprehensive plan for policy makers and practitioners if they adopt an economic development strategy along this line of thinking.

Regional entrepreneurship systems or ecosystems (REES)

Somewhat related to Figure 10.1 and the holistic approach I have suggested in the first direction, the second promising direction is the research and practice

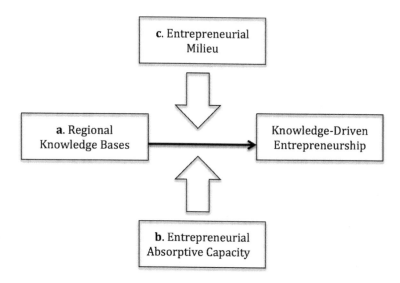

Figure 10.1 A holistic approach to the geography of knowledge spillover
    entrepreneurship

of regional entrepreneurship systems (Qian, 2010; Qian et al., 2013) or regional entrepreneurial ecosystems (Feld, 2012; Isenberg, 2011; Mack & Mayer, 2015; Spigel, 2015). While the term "ecosystem" has gained higher popularity in the literature, Stam (2014) notes that it might be more appropriate to use "system" given the man-made (versus natural) environments in this context. I consider they are the same by definition and use REES to represent both systems and ecosystems in the rest of the discussion. Different from Figure 10.1, a healthy REES does not have to involve technology entrepreneurship. Less technologically oriented entrepreneurship also contributes to regional development via employment growth, which is highly valued by policy makers or politicians, and via productivity growth by allocating market resources more efficiently (Kirzner, 1997). The system/ecosystem approach to regional entrepreneurship has a short history of only five years (Stam, 2015), but is rapidly gaining popularity thanks to joint efforts by researchers from different fields and practitioners. This research area features an interactive process in which ecosystem scholars both inform and are informed by practitioners (or ecosystem builders).

With its short history, it is not surprising that the REES research is still at the conceptualization stage. According to its name, REES adopts a systems or holistic approach to entrepreneurship, and considers all important factors that facilitate regional entrepreneurial activity, usually in an interactive way (Qian, 2010; Qian et al., 2013). That being said, the few notable REES studies (Feld, 2012; Isenberg, 2011; Mack & Mayer, 2015; Qian, 2010; Spigel, 2015) have different perspectives when conceptualizing it. The approach by Qian and his collaborators (Qian, 2010; Qian et al., 2013) is to focus on "systems." They develop a conceptual framework of REES following the way innovation systems or regional innovation systems (RIS) is introduced, e.g., by Edquist (1997), but address the lack of focus in the RIS literature on individual entrepreneurs and new firm formation. Instead of the goal of facilitating learning in RIS, they argue that the *function* of REES is to facilitate entrepreneurial activity. They particularly emphasize individual entrepreneurs as the core *component* in REES, more important than organizations and institutions widely discussed in the RIS literature. Qian et al. (2013) discuss different geographical levels of REES. Although not the core focus of their research, they identify five exogenous factors of the system, namely agglomeration, technology, universities, cultural diversity, and the quality of life.

The focus of most other notable REES studies is to identify a set of factors that can be considered as part of the ecosystem. For instance, an influential piece in the management literature (Isenberg, 2011) defines REES as six domains: policy, markets, finance, human capital, culture, and supports. Spigel (2015) expands the composition of REES into ten factors: cultural attitudes, entrepreneurial histories, networks, capital, dealmakers and mentors, talent, universities, support services and facilities, policy, and openness of market. A book written by Feld (2012), very popular among practitioners and followed by a series of research at the Kauffman Foundation, also discusses a number of components in what he calls "start-up communities" (same with REES in nature). Among them, the leading role successful, committed entrepreneurs play in mentoring other entrepreneurs is the paramount factor. There are other studies (Mack & Mayer, 2015; Stam, 2015) that build conceptual frameworks for REES in a similar way (i.e., introduce a set of factors that are relevant to regional start-up activity). Chapters in this book, though not directly using the REES framework, are more or less focused on the factors that are an important part of the ecosystem as identified in the above-mentioned literature, such as institutions (Chapter 6), networks (Chapters 3 and 9), policy (Chapter 5), entrepreneurial histories (Chapter 4), agglomeration (Chapters 7 and 8) and knowledge/technology (Chapter 2).

As an emerging area of research, there are certainly a lot of ways researchers can contribute. I see the following three tasks that are critical to the future of the REES research:

1. To make the area more coherent, it is necessary to find common grounds when conceptualizing REES. At this point, the goal of each "new" conceptualization is more or less to add "new" factors. As there are many determinants of entrepreneurship, it is easy to propose a new set of factors as a new definition of REES, and each new definition may well explain some particular REES cases. But such efforts do not contribute to the REES research as a coherent, scientific research area.

2. Discussions on REES research methodologies are needed. Given the nature of REES, it is common to see the adoption of the case study methodology, e.g., in Feld (2012) and Spigel (2015). The generalizability of particular cases is usually questioned. Qian et al. (2013) instead use a structural analysis approach to address the interactiveness of different factors. Discussions on the pros and cons of different methodologies for REES will better define this area of research and further invite more scholars into it.
3. Despite the high level of interactions between REES scholars and practitioners compared with most other research areas related to entrepreneurship, REES research has not provided workable toolkits for REES practice. It is not straightforward to convert many factors discussed in the REES literature (e.g., social capital and histories) into practice. Moreover, as many factors have been found to be important to REES, "which one(s) to start with" has not been well answered by researchers for practitioners.

Evolutionary economic geography of entrepreneurship

The third direction that shows great promise is to study entrepreneurship in the context of evolutionary economic geography (EEG), a research field that emerged during the last decade (Boschma & Frenken, 2006; Boschma & Frenken, 2011; Boschma & Martin, 2010). EEG, as defined by Boschma and Frenken (2011), studies "the spatial evolution of firms, industries, networks, cities and regions from elementary processes of the entry, growth, decline and exit of firms, and their locational behaviour" (p. 295). Indeed, entrepreneurship (from the perspective of, e.g., firm entry and growth) according to this definition is an indispensable part of EEG. Boschma and Frenken (2011) further outline three main areas in EEG: the evolution of industrial clustering, an evolutionary approach to agglomeration economies, and institutions. Additionally, they also mention knowledge networks as another area in which EEG has made some progress. Among these areas, entrepreneurship is particularly relevant to the evolution of industrial clustering, largely attributable to Klepper's work on the role of spin-offs in cluster building (Klepper, 2010).

For the purpose of studying the geography of entrepreneurship, it is more relevant to think of "EEG of entrepreneurship" instead of "entrepreneurship in EEG." As noted by Stam (2010), regional/geographical studies of entrepreneurship are dominated by a static approach. A dynamic or evolutionary approach is still much needed. Some chapters in this book contribute to understanding the dynamics of entrepreneurship (or the lack of), e.g., Chapter 3 on history dependency of entrepreneurship by Fritsch and Wyrwich, and Chapter 6 on the evolution of entrepreneurship in China as a result of institutional reforms by He, Guo, and Zhu.

Below are two research streams for future consideration using an evolutionary approach to the geography of entrepreneurship.

1. Instead of examining the role of entrepreneurship in industrial clustering, as done in EEG such as Klepper (2010), it is also necessary to understand how changes in industrial clusters may impact entrepreneurial activity. Clustering has been found to be a contributor of start-up activities (Delgado et al., 2010). But existing studies are large static. Some combinations of the product life cycle theory, clusters, and entrepreneurship will be helpful toward a dynamic approach.
2. Combined with the second research direction, it will be useful to study the evolution of regional entrepreneurship ecosystems. Despite being at its conceptual stage in the literature, REES is dynamic in nature. How it is created, grows, and declines has major implications on economic development policy. To build a vibrant entrepreneurial ecosystem, policy makers and practitioners may be particularly interested in how the ecosystem emerges and what role they can play to facilitate this process. Two authors in this book, Mack and Mayer (2015), have made some efforts on this. They discuss the changing strengths and weaknesses of different components in REES over time. However, much work is still needed.

Concluding remarks

This book is a collection of work reflecting the latest development on the research of the geography of entrepreneurship. Contributors include scholars who have been pushing forward the frontier of this research area. Indeed, this area has been flourishing and contributing tremendously to regional economic development policy. Continued work along the directions discussed above, will make further differences in the fields of entrepreneurship, geography, as well as public policy.

References

Acs, Z. (2006). How is entrepreneurship good for economic growth? *Innovations*, 1(1), 97–107.

Acs, Z. J., & Armington, C. (2004). The impact of geographic differences in human capital on service firm formation rates. *Journal of Urban Economics*, 56(2), 244–278.

Acs, Z. J., & Armington, C. (2006). *Entrepreneurship, geography, and American economic growth*. New York: Cambridge University Press.

Acs, Z. J., Desai, S., & Hessels, J. (2008). Entrepreneurship, economic development and institutions. *Small Business Economics*, 31(3), 219–234.

Acs, Z. J., Braunerhjelm, P., Audretsch, D. B., & Carlsson, B. (2009). The knowledge spillover theory of entrepreneurship. *Small Business Economics*, 32(1), 15–30.

Aernoudt, R. (2004). Incubators: tool for entrepreneurship? *Small Business Economics*, 23(2), 127–135.

Andersson, M., & Koster, S. (2011). Sources of persistence in regional start-up rates: Evidence from Sweden. *Journal of Economic Geography*, 11(1), 179–201.

Asheim, B. T., & Coenen, L. (2005). Knowledge bases and regional innovation systems: Comparing Nordic clusters. *Research Policy*, 34(8), 1173–1190.

Asheim, B. T., Boschma, R., & Cooke, P. (2011). Constructing regional advantage: Platform policies based on related variety and differentiated knowledge bases. *Regional Studies*, 45(7), 893–904.

Asheim, B., Coenen, L., & Vang, J. (2007). Face-to-face, buzz, and knowledge bases: Sociospatial implications for learning, innovation, and innovation policy. *Environment and Planning C*, 25(5), 655.

Audretsch, D. B., & Lehmann, E. E. (2005). Does the knowledge spillover theory of entrepreneurship hold for regions? *Research Policy*, 34(8), 1191–1202.

Audretsch, D., Dohse, D., & Niebuhr, A. (2010). Cultural diversity and entrepreneurship: A regional analysis for Germany. *The Annals of Regional Science*, 45(1), 55–85.

Audretsch, D. B., Leyden, D. P., & Link, A. N. (2013). Regional appropriation of university-based knowledge and technology for economic development. *Economic Development Quarterly*, 27(1), 56–61.

Boschma, R. A., & Frenken, K. (2006). Why is economic geography not an evolutionary science? Towards an evolutionary economic geography. *Journal of Economic Geography*, 6(3), 273–302.

Boschma, R., & Frenken, K. (2011). The emerging empirics of evolutionary economic geography. *Journal of Economic Geography*, 11(2), 295–307.

Boschma, R., & Martin, R. (Eds.) (2010). *The handbook of evolutionary economic geography*. Cheltenham: Edward Elgar Publishing.

Capello, R. (2013). Territorial patterns of innovation, in Capello, R. & Lenzi, C. (Eds.) *Territorial patterns of innovation: An inquiry on the knowledge economy in European regions*, 129–150. New York: Routledge.

Delgado, M., Porter, M. E., & Stern, S. (2010). Clusters and entrepreneurship. *Journal of Economic Geography*, 10(4), 495–518.

Edquist, C. (Ed.). (1997). *Systems of innovation: Technologies, institutions and organizations*. London: Pinter.

Feld, B. (2012). *Startup communities: Building an entrepreneurial ecosystem in your city*. Hoboken: John Wiley & Sons.

Feldman, M., & Zoller, T. D. (2012). Dealmakers in place: Social capital connections in regional entrepreneurial economies. *Regional Studies*, 46(1), 23–37.

Fritsch, M., & Wyrwich, M. (2014). The long persistence of regional levels of entrepreneurship: Germany, 1925–2005. *Regional Studies*, 48(6), 955–973.

Haltiwanger, J., Jarmin, R. S., & Miranda, J. (2013). Who creates jobs? Small versus large versus young. *Review of Economics and Statistics*, 95(2), 347–361.

Isenberg, D. (2011). The entrepreneurship ecosystem strategy as a new paradigm for economic policy: Principles for cultivating entrepreneurship. *Presentation at the Institute of International and European Affairs*.

Kirzner, I. M. (1997). Entrepreneurial discovery and the competitive market process: An Austrian approach. *Journal of Economic Literature*, 60–85.

Klepper, S. (2010). The origin and growth of industry clusters: The making of Silicon Valley and Detroit. *Journal of Urban Economics*, 67(1), 15–32.

Lee, S. Y., Florida, R., & Acs, Z. (2004). Creativity and entrepreneurship: A regional analysis of new firm formation. *Regional Studies*, 38(8), 879–891.

Mack, E. A. (2014). Businesses and the need for speed: The impact of broadband speed on business presence. *Telematics and Informatics*, 31(4), 617–627.

Mack, E. A., & Mayer, H. (2015). The evolutionary dynamics of entrepreneurial ecosystems. *Urban Studies*, 0042098015586547.

Malecki, E. J. (1993). Entrepreneurship in regional and local development. *International Regional Science Review*, 16(1–2), 119–153.

Minniti, M. (2008). The role of government policy on entrepreneurial activity: Productive, unproductive, or destructive? *Entrepreneurship Theory and Practice*, 32(5), 779–790.

Nijkamp, P. (2003). Entrepreneurship in a modern network economy. *Regional Studies*, 37(4), 395–405.

Qian, H. (2010). *Regional systems of entrepreneurship*. Doctoral dissertation at George Mason University.

Qian, H. (2013). Diversity versus tolerance: The social drivers of innovation and entrepreneurship in US cities. *Urban Studies*, 50(13), 2718–2735.

Qian, H., & Acs, Z. J. (2013). An absorptive capacity theory of knowledge spillover entrepreneurship. *Small Business Economics*, 40(2), 185–197.

Qian, H., & Haynes, K. E. (2014). Beyond innovation: The Small Business Innovation Research program as entrepreneurship policy. *The Journal of Technology Transfer*, 39(4), 524–543.

Qian, H., Acs, Z. J., & Stough, R. R. (2013). Regional systems of entrepreneurship: The nexus of human capital, knowledge and new firm formation. *Journal of Economic Geography*, 13(4), 559–587.

Reynolds, P., Storey, D. J., & Westhead, P. (1994). Cross-national comparisons of the variation in new firm formation rates. *Regional Studies*, 28(4), 443–456.

Schumpeter, J. A. (1934). *The theory of economic development: An inquiry into profits, capital, credit, interest, and the business cycle*. Cambridge, MA: Transaction Publishers.

Smith, H. L., & Bagchi-Sen, S. (2012). The research university, entrepreneurship and regional development: Research propositions and current evidence. *Entrepreneurship & Regional Development*, 24(5–6), 383–404.

Spigel, B. (2015). The relational organization of entrepreneurial ecosystems. *Entrepreneurship Theory and Practice*, DOI: 10.1111/etap.12167.

Stam, E. (2010). Entrepreneurship, evolution and geography, in Boschma, R. & Martin, R. (Eds.) *The handbook of evolutionary economic geography*, 139–161. Cheltenham: Edward Elgar.

Stam, E. (2014). The Dutch entrepreneurial ecosystem. Available at SSRN 2473475.

Stam, E. (2015). Entrepreneurial ecosystems and regional policy: A sympathetic critique. *European Planning Studies*, 23(9), 1759–1769.

Welter, F. (2011). Contextualizing entrepreneurship: Conceptual challenges and ways forward. *Entrepreneurship Theory and Practice*, 35(1), 165–184.

Westlund, H., & Bolton, R. (2003). Local social capital and entrepreneurship. *Small Business Economics*, 21(2), 77–113.

Index

Taylor & Francis eBooks

Helping you to choose the right eBooks for your Library

Add Routledge titles to your library's digital collection today. Taylor and Francis ebooks contains over 50,000 titles in the Humanities, Social Sciences, Behavioural Sciences, Built Environment and Law.

Choose from a range of subject packages or create your own!

Benefits for you

» Free MARC records
» COUNTER-compliant usage statistics
» Flexible purchase and pricing options
» All titles DRM-free.

Free Trials Available
We offer free trials to qualifying academic, corporate and government customers.

Benefits for your user

» Off-site, anytime access via Athens or referring URL
» Print or copy pages or chapters
» Full content search
» Bookmark, highlight and annotate text
» Access to thousands of pages of quality research at the click of a button.

eCollections – Choose from over 30 subject eCollections, including:

Archaeology	Language Learning
Architecture	Law
Asian Studies	Literature
Business & Management	Media & Communication
Classical Studies	Middle East Studies
Construction	Music
Creative & Media Arts	Philosophy
Criminology & Criminal Justice	Planning
Economics	Politics
Education	Psychology & Mental Health
Energy	Religion
Engineering	Security
English Language & Linguistics	Social Work
Environment & Sustainability	Sociology
Geography	Sport
Health Studies	Theatre & Performance
History	Tourism, Hospitality & Events

For more information, pricing enquiries or to order a free trial, please contact your local sales team:
www.tandfebooks.com/page/sales

Printed and bound by CPI Group (UK) Ltd, Croydon, CR0 4YY

22/10/2024

01777628-0012